Northern Bee Books
Scout Bottom Farm
Mytholmroyd
Hebden Bridge
HX7 5JS (UK)

www.northernbeebooks.co.uk

Tel: 01422 882751

© 2019

ISBN: 978-1-912271-45-0

The Photo of the "Honey Bee Swarm on Mailbox" on Front Cover is Courtesy of Harry Hillard, New Jersey.

Bees
and Man

Bumble Bees ● Carpenter Bees ● Honey Bees ● Hornets ● Yellowjackets

William Michael Hood, PhD

Preface

Bees have provided man a rich source of experiences that have been shared down through the ages. My work in support of the beekeeping industry for over 30 years offered me many great opportunities to meet some interesting individuals who came from diverse backgrounds. I met these people simply because we had one common interest and that was interacting or working with bees which produced some bizarre and unpredictable situations that I have chosen to record and share through this book.

The word "bee" is normally used to refer to honey bees, but in a broader context of the term, bees will include other hymenopterous insects found in the super families Apoidea and Vespoidea. The superfamily Apoidea includes honey bees, bumble bees, and carpenter bees, whereas the superfamily Vespoidea includes wasps, hornets, and yellowjackets. You will find stories of all of the above listed insects and others in this book.

Down through recorded history, bees have provided man a source of excitement and fascination. They have provided man a source of defense in wartime as well as a food source in the form of honey. Honey bees have provided man another benefit in the name of apitherapy which is defined as the use of bee venom and other beehive products to treat various human medical conditions. Bees are well respected by man because of their ability to defend their colonies and themselves. However, man has learned over the years to manage honey bees and take advantage of their ability to produce enormous amounts of honey and to value them as pollinators of fruits, vegetables, and plants for wildlife.

My hope is that you will read this book and enjoy some of the stories that I have lived out during my work as professor, bee specialist, and state apiarist at Clemson University. My intent is not to offend anyone, but to tell the stories as I perceived them to play out from some interactions with other individuals that turned from ordinary circumstances to develop into some very entertaining stories. Some readers especially novice beekeepers may learn many valuable lessons from reading this book. However, I feel that everyone, experienced beekeepers as well as non-beekeepers, will enjoy some of the varied and sometimes bizarre stories that, by the way, are true. Most of the stories are from my own experiences, but a couple were made available or shared with me by reliable sources who I'm convinced were telling the truth.

So, my hope is that you will read and enjoy.

Wm. Michael Hood, PhD
Emeritus Professor of Entomology
Emeritus College
Clemson University
Clemson, South Carolina

Acknowledgements

A special thanks to Paul Boone for permission to use his photos of plants, trees, and impressive images of pollen grains of various plants and trees. Thanks to others who allowed me to use their photos including: Diane Biering, Wayne P. Armstrong; Betterbee, Inc.; Jeffrey Lotz; Laurence Cutts; Eric Mills; David Christopher; Clemson University Extension Program, Florida Dept. of Plant Industry; Animaltrappingremoval.com; City of Chesapeake, VA; Glorybee; Wildflower Meadows, Loudoun Beekeepers Association; Nelson Paint Company, Renae Ausburn, Lissa Gotwals, John R. Steedly, Wyatt Mangum; Harry Hillard; Pixabay.com; Mike Bentley; Jimmy Tinsley; Rick Glover and Bob Bellinger.

The author wishes to thank his wife, Kathy, for her support while writing this book and to thank her for reviewing the manuscript and making valuable suggestions. Thanks to my daughter, Lindsey, for her timely suggestions.

Contents

Odd Bee Stories

Honey Bee Flight Pattern Over School Parking Lot

The chairman of a local school board called me one day with a request to assist him with an issue involving an uncooperative beekeeper. This beekeeper, whose property was adjacent to one of his elementary schools and within city limits, had about 75 beehives in his backyard which were visible to the public.

Unfortunately, the flight pattern of the foraging honey bees from these colonies took them right over the school parking lot. During major spring nectar flows in April and May, the bees were often overloaded on their way home with nectar and pollen and sometimes lost part of their load which fell to the ground or onto the school employee's vehicles.

This happened to be a year when there must have been a bumper high nectar flow because the bee droppings on the vehicles had reached an all time high and the school principal had received many complaints. Normally the bee droppings were washed away by rain or were easily washed off by the employees. However, apparently this year the employees had too much of the "bee poop" as they called it. Some employees had mentioned that the staining had become so bad that it appeared they may need to have their vehicles painted.

The school board chairman requested that I contact the beekeeper to see if I could help with the issue because he had gotten little cooperation from the gentleman. He asked if

Fig. 1. Honey bee droppings on hood of vehicle (Photo by Wayne P. Armstrong, San Marcos, CA).

I could talk to the beekeeper and get him to reduce his number of honey bee colonies down to a manageable level, say ten colonies. I agreed reluctantly to assist in this issue and arranged to meet with the beekeeper and discuss this matter.

I met him for lunch the next day and the beekeeper admitted that he was well aware of all the recent complaints. He informed me that he lived there well before the school was built adjacent to his property. He also informed me that when the school was built, the "school administration" took part of his property and he was very displeased with the outcome of the settlement. In a way of getting back at the school administrators, he was reluctant to cooperate with the school officials in anyway.

Next, we discussed the issue of the school board chairman's suggestion of reducing the number of honey bee colonies down to a manageable level of approximately ten. A sheep-ish grin came over the beekeeper's face as he admitted that the parasitic mites had already reduced his operation down to less than ten live bee colonies. He noted that the other 65 or so beehives were just empty boxes that he kept painted and visible just as an annoyance to the public, especially school officials and administrators. They never bothered to inves-tigate and discover that there were no bees in most of the beehives.

I reported back to the school board chairman the next day and told him that the beekeeper agreed that he would only manage ten or fewer bee colonies in his backyard. The chairman seemed to be happy with the outcome of events. I never heard back on this matter, but I suspect mother nature helped as the nectar flow diminished over the next few days and apparently the "bee poop" stopped raining down on the cars in the school parking lot.

An Uncommon Beehive

Part of my initial responsibilities at the university included honey bee regulatory work and my official title was known as "State Apiarist" or commonly known as state bee inspector. In the late 1980s, we were very concerned about a new parasitic honey bee pest, the varroa mite, that had been found infesting honey bee colonies in many other states in the US and we knew that we would eventually find the pest in South Carolina.

We were in the process of conducting a statewide varroa mite survey including some remote locations in order to provide early detection of the pest which might occur in some unexpected places. One method of finding beekeepers in remote locations was to talk to other beekeepers and "pick their brains" for names of beekeepers who might not have been included in our past surveys. One beekeeper who I knew well informed me of a beekeeper who lived "back in sticks" in northern Pickens County.

I drove up in the beekeeper's yard one day and introduced myself as the state bee inspector and that I was there to conduct a varroa mite survey in some of his bee colonies. I could tell that he was totally unfamiliar with the term "varroa mite" and initially it appeared he might not be interested in having his bee colonies inspected, sort of a feeling of government intrusion. After much discussion of the dangers of this pest, he reluctantly agreed to cooperate with my survey.

During our discussions, I noticed that he had some of the roughest and unusual looking beehives in his yard that I'd ever seen. Most of the equipment looked to be homemade of any type wood that he could find. However, I have found over the years that honey bees could care less about the quality or appearance of the beehive structure as long as the inside of the structure was big enough, had dry conditions inside, and had a small entrance that could be protected from predators. This beekeeper must have had about 40 colonies that I could see that were visible, mostly in his front yard.

While walking through his beeyard to select colonies that I could inspect, I noticed he was reluctant to show me some colonies on the backside of his lot. My first impression was that there was something there that he did not want me to see. But I insisted that the survey needed to represent his entire operation, so we walked over to these additional colonies and there I saw what he did not want me to see. He had a bathroom toilet bowl set up with a honey super placed on top.

Fig. 2. The white toilet bowl beehive.

I learned from him that the story went like this: One spring the beekeeper discovered that a swarm of honey bees had settled and set up housekeeping in an abandoned toilet that was located in a trash heap on his property. Rather than attempt to remove the bees from the toilet and risk killing the colony, he decided to relocate the toilet bowl with bees and place it among his other colonies. He claimed to have harvested honey from the colony in the past. I'm certain that I would not want to consume the honey harvested from his toilet bowl beehive, however the bees appeared to be quite content in their nice white abode or should I say commode.

Lesson learned. I can assure you that I had never seen nor heard of such a beehive, but this is another example of an odd occurrence that you may run across when working with bees and man. The inside dimensions of the toilet bowl were much too small to house the brood area of a normal honey bee colony. Research indicates that the ideal size of a brood nest should be about 40 liters (10.5 gal.) with an entrance size of 15 cm² (2.3 in²). So, I'm convinced that this colony swarmed every spring due to the small volume of the brood nest area. This is certainly not a practical way to manage a colony of honey bees, mainly because the colony could not be inspected or treated for maladies such as disease and pests.

The Chimney Beehive

The late Billy Bostic (1936-2011), of Starr, South Carolina, invited me to come and visit him to see his beekeeping operation. Billy had always been very helpful to me by his willingness to follow up on bee swarm calls or to remove bee colonies from structures in southern Anderson County. One day, I took him up on his invitation and paid him a visit.

As soon as I exited my car, Billy met me and showed me his very impressive honey house. He had purchased most of the equipment from well-known and long-time Anderson bee-keeper, Tom Chapman. Tom had become allergic to bee venom later in life and, upon strong recommendation from his doctor, he had given up beekeeping.

Billy led me around to his backyard apiary and showed me one of his favorite beehive pro-ductions. Story goes that Billy received a phone call from a family who was tearing down an old house on their property and they discovered that their still-standing chimney had a colony of honey bees inside that had taken up residence. They did not want to kill the bees, so Billy got the call.

Billy told the family that he would gladly come over and take a look at the bees in the chimney and would try to salvage them, if possible. Billy arrived on scene and noticed very quickly that the bees could not be removed from the chimney without high risk of killing them. So, Billy did the next best thing and transported the upper half of the chimney containing the bee colony to his home and placed it among his other colonies. He placed a honey super on top of the chimney and claimed to have produced surplus honey from it some years.

Fig. 3. Billy's Chimney Honey Bee Colony.

NOTE: As with the toilet bowl beehive story above, this is not a practical method of managing a colony of honey bees because the beekeeper cannot open the colony and inspect for queen status, brood disease, parasitic mites, and food/pollen stores.

<u>Lesson learned</u>. In South Carolina, it is not illegal to keep honey bees in a structure of this nature or in a hollow of a tree that has been cut from what is known as a bee tree. However, this kind of beekeeping is illegal in many other states because the bees cannot be properly managed for disease and pests. In other words, the brood rearing area of the colony cannot be taken apart and inspected. In our state, we highly discourage beekeepers from keeping honey bees in structures of this nature because the colony could become a source of contamination to other nearby colonies.

Woody Mites in York County

The late Tony Hepp who lived in York County was a commercial beekeeper who continued to keep honey bees later in life. He was near the age of 80 at the time of this story. Like many senior aged beekeepers, Mr. Hepp was challenged to learn how to successfully manage his honey bees with the onset of parasitic mite infestations that began in the mid-1980s in South Carolina.

In 1984, tracheal mites were the first devastating honey bee mite species discovered in the US. These mites were microscopic in size and could not be seen with the unaided eye. Many beekeepers did not pay attention to something they could not see and experienced great losses of their honey bee colonies from the effects of this pest for a few years.

Mr. Hepp was one of many South Carolina beekeepers who lost over half their colonies to tracheal mites in a very short period of time. He had recently submitted bee samples from some of his weak colonies to the USDA Honey Bee Disease and Pest Diagnostic Lab in Beltsville, Maryland, for testing.

During this same period of time, I was very busy traveling to various local beekeeper association meetings giving training sessions on how to manage their colonies when tracheal mites were present. Tracheal mites were a hot topic at local beekeeper association meetings because their bee colonies were being decimated by this mite.

One such meeting was held by the York County Beekeepers Association where there was a large turnout of beekeepers in attendance. I gave my normal tracheal mite slide presentation on the life cycle of tracheal mites, how to identify them, and how to control the pest. After my 45-minute presentation on tracheal mites, I asked for questions. As it turned out, an older gentleman sitting on the front row was Tony Hepp, who I had never met. I noticed that he was challenged to hear much of my presentation, but he had a question at the end. He asked, "Dr. Hood, have you ever heard of the woody mite?"

I pondered for a little while on that question hoping that someone in the audience knew what he was talking about. Finally, it dawned on me that Mr. Hepp was referring to the species name of tracheal mites. I learned later that he had recently received a report from the USDA lab on his bee samples which he had submitted earlier. The report had come back as "positive for *Acarapis woodi*" which is the scientific name or Latin name for tracheal mites. Mr. Hepp apparently did not care to pronounce the genus name *Acarapis*, but he readily pronounced the species name "*woodi*." So, that is where his term "woody mites" came from. I was little bit surprised, that after he had listened to my 45-minute presentation, he did not connect the scientific name with the common name "tracheal mites."

Lesson learned. From that day on, I made sure to cover both the common and scientific names of tracheal mites in all future training sessions. Although Mr. Hepp was challenged by the parasitic mites later in life, he was known as an excellent beekeeper during his prime years. His beekeeping specialty was providing local businesses multi-jar packs of varietal honeys of different colors as gifts to their employees or significant clients on special occasions. According to Mr. Hepp, he did quite well with repeat sales to many of his regular customers. Other beekeepers may consider this unique opportunity as a niche market for their honey sales.

There Are Honey Bees in My Couch

I received a call one day in my office from a lady in Seneca, South Carolina, who said, "I have honey bees in my couch. Do you think you could help me?" I had her carefully repeat what she had said and sure enough there was a honey bee colony inside an old couch that she had placed outside behind her house. She claimed that some young boys in her neighborhood had been throwing sticks at the bees and that she feared the boys were going to get stung.

Fig. 4. Lady observing a colony of honey bees nesting in her couch. Note the sticks laying on the left side of the couch the boys had been throwing at the bees.

I told her that I knew a nearby beekeeper, Clyde McCall, who lived in Seneca and that he should be able to remove the bee colony for her. I called Clyde and met him at her house that same day because I wanted to take some photos of such a bizarre bee story. Sure enough, a swarm of honey bees had entered the front side of the couch likely the spring before and they appeared to be pretty content and comfortable in their new home.

Several conditions contributed to this acceptable home for the honey bees. The couch had been placed behind her house which faced south, providing plenty of warmth in winter and a south facing entrance which bees prefer. The back of the couch had a void large enough to support the colony and no pest or predator had disturbed them enough to make them leave.

Clyde and I turned the couch over and cut open the back of the couch. We discovered several layers of beautiful comb filled with bees, honey, and bee brood. Clyde cut most of the comb out and fastened it to hive frames placing them into a hive body and transported them that evening to his home yard. The lady was happy to see the bees leave and Clyde was happy to get a new colony of honey bees.

Fig. 5. Beekeeper Clyde McCall examining the exposed bee colony.

Lessons learned. This story is good example of a beekeeper being a good ambassador for the beekeeping industry. If Clyde had not rescued this colony of honey bees, the homeowner may have called a pest control company and you can imagine the unfortunate demise of the honey bees. Sometimes beekeepers are faced with honey bee colony removals that are fairly simple as in this case. When removing bee colonies of this nature, beekeepers need to keep in mind a few basic guidelines such as using minimum smoke in order to find the queen which may evade the beekeeper when too much smoke is used. The beekeeper should take along an empty hive body with bottom, top cover, and several empty hive frames that have no foundation. Find the queen, if possible, and store her temporarily in a queen cage and place her into the empty hive body. Carefully cut some of the brood comb to proper dimensions and secure it into empty hive frames with rubber bands or string. The bees will secure the comb further. Place the secured frames of brood into the hive body and carefully release the caged queen onto her brood. Place the top cover over the hive body, then remove all other comb from the original colony and shake off the bees near the hive body entrance. Position the entrance of the hive body as close to the original colony entrance as possible to catch incoming foragers and leave it there for several hours. Note: There are other methods of how to remove unwanted honey bee colonies from structures, but these are a few recommendations that I have learned over the years.

Making Life Better with Honey Bees

I do not recall Mr. White's first name, but I do remember him very well. At the time I first met him, he was in his mid-eighties and managed about 15 honey bee colonies on his small rural farm in Greenwood County, South Carolina. On my first visit to meet Mr. White, the Greenwood County Extension Agricultural Agent joined me. A couple of his honey bee colonies were not productive and would not build up in spring. I was able to diagnose the problem as European Foulbrood which is a bacterial stress disease that is not fatal, but it does kill a couple of bee brood cycles causing the colony to be unproductive for the remainder of the year. A colony will normally overcome the effects of this brood disease by summer, however it has a high chance of reoccurring the next spring if left unchecked.

Mr. White's beeyard was on the backside of his property and you could tell that he truly enjoyed his time working with his honey bees. He had a rocking chair strategically located in the center of his beeyard, so that he could sit and spend time at his leisure watching his honey bees work.

A couple of years passed and I received another call from the same county agent requesting that I come down and inspect Mr. White's honey bee colonies again. Some of his colonies seemed to be unhealthy. I agreed to pay Mr. White another visit. I drove to Greenwood one day the next week and knocked on the White's front door. Mrs. White came to the door and informed me that he was probably spending time with his honey bees. I knew what she had in mind and sure enough when I walked down to the beeyard, there was Mr. White sitting in his rocking chair watching his bees work. It happened to be a beautiful spring day and conditions were just perfect for watching bees, according to Mr. White. If my memory serves me correctly, I diagnosed the problem as European foul brood again, but the joy of seeing him rocking in his beeyard was worth the trip to Greenwood and overshadowed any other issues.

Lessons learned. Beekeeping is an important activity that includes more than just producing honey and providing bees to pollinate our fruits and vegetables. Just as important, beekeeping provides a person time to unwind from many of life's stresses and to enjoy working with one of God's greatest creations, the honey bee. In some beekeeping circles, it is common knowledge that when a beekeeper dies that another beekeeper should visit the beeyards of the deceased and convey to the bees that their keeper has passed. AMEN!

Fig. 6. Honey bee swarm looking for its long-gone keeper?

Count All the Bees in the Hive

Sometimes our honey bee research schedule required us to work with honey bees during cool or inclement weather and our protocol often called for us to completely dismantle colonies of bees to collect colony measurements such as number of adult bees in the beehive. The process went like this. After closing up the beehive and confining the bees at night, we weighed the total beehive and bees on a field balance the next morning. After capturing and safely confining the queen, we removed each frame from the hive and shook or brushed the bees off with many of the bees taking flight. We placed a dummy hive body in the exact same location of the parent hive so the bees would have a box to go into temporarily. After removing all the bees from the frames and boxes, we reweighed just the equipment and brood without the bees. The original weight minus the final equipment weight gave the total weight of the adult bees. After weighing several individual bees, we obtained an average bee weight that we divided into the total bee weight which gave us the number of bees in the colony. Then, we removed the dummy hive box, shook the bees into the reassembled original equipment in the same location and introduced the queen back into the colony. All colonies in an apiary were processed in the same manner to estimate the adult bee population for the study.

This unnatural ordeal, often conducted during times of cool weather, resulted in many bees taking flight and seeking a warm surface to settle. The warmest nearby surface was created by our bodies. Often the bees settled between our legs which created a very sensitive situation or the bees settled overhead on our bee helmet or on our backside. By the time we finished the job, we looked like a walking bee swarm weighed down by several thousand bees. It was these times that I remembered that we were especially glad to have worn bee protection. We carefully brushed the bees from our crotch and then jumped up and down a few times to dislodge the other bees, which by then we had re-assembled their

Fig. 7. Wow! That's a lot of bees on my backside.

hive in the original location. The bees returned to their hive in the beeyard eventually and everyone was happy again. To say the least, this procedure was tough on the bee colonies, but it was an event that gave us needed data for our research.

At other times during our research, we were conducting experiments on small hive beetles which were a new hive pest that entered South Carolina in 1996. On this occasion, we were conducting total hive beetle counts requiring us to shake all the bees and beetles off the frames and boxes onto a white table. We carefully brushed the bees to the side and counted the adult beetles. This proved to be a frantic ordeal that left the bees completely confused, so again many of the bees took flight and landed on the warmest object available which happened to be our backsides.

Lesson learned. Bee scientists sometimes do unusual things which do not make a lot of sense to the public. You might say that there has to be a better way. There might be, but if we always knew what we were doing, it would not be called research.

Fig. 8. Now bees get back in your beehive.

Bananas for Lunch

One summer, I had a home-schooled student who interned with me for a few weeks on a special program for young high school scholars. Andy was a rising junior and he was a delightful young man to work with. One of his projects while working with me was investigating the effectiveness of a trap that we designed for controlling wax moths in the beeyard.

Wax moths are a pest insect that invades and lays eggs in beehives in late evening. A strong or healthy colony will remove and kill wax moths in the larval stage. However, wax moth larvae can result in major damage to a honey bee colony if it is stressed from disease, starvation, or some other malady. Our goal was to create a trap that would attract and kill the wax moth adults before they entered the beehives to lay eggs.

The day came to begin Andy's field study, so I traveled to the local grocer and purchased the ingredients for the traps. For each trap, the project called for one banana peel, one cup of white vinegar, one cup of sugar and about two cups of water placed in a 2-liter plastic soda bottle for wax moth control. The plastic bottle had a hole cut about 1.5 inches in diameter just below the shoulder which allowed moth entry. I returned with the ingredients for the project just before lunch that day and told Andy that he was welcome to eat the bananas with his lunch and that I'd return in about an hour.

Fig. 9. Wax moth trap loaded with ingredients.

Upon my return, Andy claimed that he was unable to eat all the bananas for there were more than he could eat with his lunch. However, it was evident that he had eaten 4-5 of the bananas. It appeared to me that he had assumed that his consumption of the bananas was part of the research project. I'm not sure what all those bananas did to his digestive system that day, but I learned a good lesson to make very clear your intentions and expectations to young students when doing a research project.

Fig. 10. Andy setting up traps in our beeyard for wax moth control investigations.

Lessons learned. As I understand, this young man went on to college and eventually earned his PhD in Entomology. I learned to never underestimate your experiences and influence when working with children and youth. We learned from these preliminary investigations that the wax moth trap described in this story should not be used in an apiary during a nectar dearth period, because honey bees may be attracted to the trap that holds a sugary solution and they may die. Otherwise, we found that wax moth adults will enter the trap and die during other times of the year.

Ultimate Security System: Beehives

During my first few years working at Clemson University, not only was I the extension bee specialist and state bee inspector, but I also had the regulatory job of inspecting plant and tree nurseries and greenhouse plants in seven upstate counties for pests and disease. On one of my annual inspection visits to a greenhouse grower's operation, the owner told me that he had experienced some vandalism problems to his plants inside his greenhouse in the past. He suspected that it was some local kids "who were up to no good." The intruders did not steal anything, but their activities resulted in turned over pots, damaged plants, and generally made a mess of things.

The grower was also a beekeeper and he knew that most people are fearful of stinging insects. So, he placed a hive of honey bees on each side of his greenhouse just inside the fence. He claimed the honey bees had solved his vandalism problem and he had not been visited by the vandals again.

Fig. 11. No more greenhouse vandalism with this security system in place.

A Call from the Athletic Department

One morning in early August, I was sitting at my desk and received a call from the head athletic trainer for the Clemson University football team. He said that they were having a major problem with honey bees at their practice field on campus. He claimed the bees were going after their Gatorade® station causing quite a disruption.

I assured him that these were yellowjackets which are often a problem around any sugary liquid material this time of year. The yellowjacket population builds from early summer and peaks about August and they can become quite a nuisance, especially around picnic tables and trash cans.

The head trainer said that he understood, but he was very sure that these were honey bees and not yellowjackets. He even gave me a good description of the bees and sure enough it sounded like these were honey bees. I told him that I'd be right down to the practice field and survey the situation.

I drove my university pickup truck to the practice field and was allowed to enter a security gate, without being questioned. The security guard must have been expecting me. I got out of my truck and walked across the field toward a tent which I figured was the Gatorade® station. As I walked over, I noticed what looked like over a hundred football players and coaches doing what football teams do in practice. I imagined that some of those big 300 lb. lineman were extremely scared of those tiny honey bees.

As I approached the tent, I was enthusiastically met by the head trainer who took me to the bee problem area. It was apparent that he was happy to see me show up. He described the problem again and showed me the bees which were indeed honey bees. I could see maybe 40-50 honey bees flying under the tent or trying to get inside the approximate 5-gallon containers of Gatorade®. The trainers were filling small cups with the Gatorade and placing them onto trays and delivering them to the players during breaks in their practice session. The bees were flying around the trays and occasionally a bee would fall into a filled cup according to the head trainer, making the other trainers and football players very uncomfortable.

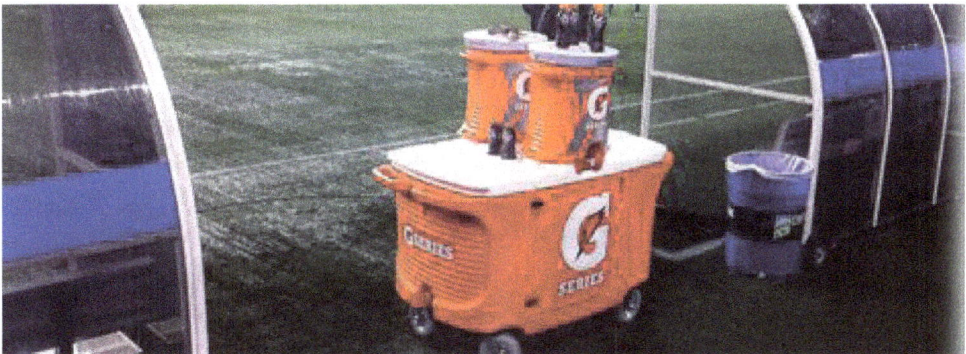

Fig. 12. Honey bees like Gatorade® too, especially in hot weather.

The head trainer said that he needed a solution to the bee problem and the sooner the better. He noted that he was aware that honey bees are very beneficial and that he did not want to kill them. Then he asked me, "Where are all these honey bees coming from?" I mentioned to him that honey bees sometimes forage for up to two miles from their home location and that it would be difficult to say for sure where the bees were nesting. I did mention that I had some bee colonies on campus.

After much discussion, I asked him the sugar content of the Gatorade® which he responded was a certain amount. I had an idea that since the bees had established a foraging pattern to fly toward the practice field, it would be very difficult to get the bees to fly in another direction.

Honey bees have a way of communicating to other foraging bees the direction and distance to a good nectar source or other food source. This is called the "waggle dance" that is performed on the hive comb by some of the incoming foraging bees. The other bees can interpret the dance language and head off in the direction of the food source, which can be from a tree or plant or it can be from a source such as Gatorade® in this case. The bees have a tremendous sense of the sugar concentration level of nectar or other potential food and will switch eventually to another source of greater sugar content.

My recommendation to the head trainer was to set up a separate nearby alternate dummy station with Gatorade® made with twice the sugar content. The bees would gradually switch to the new food source of higher sugar content and they would abandon his very active tent station. He listened very closely to my recommendation and seemed to agree that this idea made sense to him also.

I walked away feeling pretty good that this bee problem would soon be solved. I expected a call back from the head trainer, but he never called. I guess he was a busy man that time of year, so I should have followed up, but I didn't. I am certain that the bees were coming from my 10-colony apiary which was located less than a half mile from the practice field. The bee colonies were placed there for the purpose of helping pollinate vegetables on the university's organic vegetable farm. This bee foraging activity to the football practice field was certainly beneficial to my nearby colonies. The sugar content was good for the bee's future needs, but the water may have been just as important to help the bees cool their hives, since this was a hot and dry period of the year.

Lesson learned. This story confirmed the fact that mid-August is a possible time of nectar dearth in the Clemson area for the honey bees to be attracted to such a weak sugar solution like Gatorade®. If you think the bees made Gatorade® honey from the experience, you are wrong. Honey bees cannot collect sugar of this nature and convert it to honey. It will just be a sugary tasting liquid which the bees can store in the hive for their food consumption needs in winter or use it to cool the hive. Honey is produced only from plant or tree nectar.

Bee Go®

Early on in my research career, I found a product that I did not care for that was used by many beekeepers to rid honey bees from their honey supers at harvest time. The product name was "Bee Go®" and the active ingredient was Butyric Anhydride which gives off an odor that smells just like human vomit, to me. It is a great product if you can put up with the offensive odor. I reluctantly used this amazing product mainly because it was the only thing that would do the job quickly and it left behind no lingering odor in the honey or equipment. Although when using the product, the odor somehow makes its way into the fabric of your clothing, but it could be removed easily by washing.

Fig. 13. Pint bottle of Bee Go®
(Photo Courtesy of Betterbee, Inc.)

My research counterpart at the University of Georgia, Dr. Keith Delaplane, and I once had a project that required the measurement of honey production in honey bee colonies and we used Bee Go® to clear the bees from their honey stores. After using this product for a full morning of honey removal from several colonies, we visited a nearby fast food restaurant, which happened to be very busy that day. When we walked up to the waiting line to place our food order, I noticed the line sort of emptied and everyone moved to the other line. You guessed it, Bee Go® did the job and we moved right up to the register with no wait. I learned that if you need some space or privacy, Bee Go® will do the job!

Lesson learned. Bee Go® is an excellent product to remove honey bees from their honey stores, however some beekeepers including myself do not care for the offensive odor of the product. As an alternative, Bee Quick® and possibly other products are now available that have a more pleasant odor.

Honey Bees Get an Early Wakeup

Joe Gagnon, a Fox Carolina Early Morning News Reporter for TV Station 21 in Greenville, South Carolina, called me one day and requested a visit to Clemson to do a story on honey bees. Honey bees had been a regular topic in recent newspaper columns and Joe just had to get a live story on bees. We set a date and time to meet at the Cherry Farm where I had several honey bee colonies. The time of day for his visit was 5:30 AM and his news report was live. I mentioned that honey bees will not be flying that time of day, since it will be dark and that they may not like to be handled that early in the morning. That did not deter Joe, who was adamant about getting his story.

I arrived at the Cherry Farm at the appointed time of 5:30 AM that morning and sure enough Joe and his camera lady were all setup for the show with their tall TV antenna and glaring lights setup right near my beeyard. I told Joe that I could never remember opening a honey bee colony that early in the morning and that we might be in for a surprise.

I'll have to say that newsman Joe was a brave soul after all the warnings I had given him. I suited him and his camera lady up and we headed over to the bee colonies where they had earlier set up some special lighting for the show. I showed him my bee smoker and demonstrated how to light it and explained its purpose for his TV audience. It was still pitch dark except for the artificial lighting. After smoking one of my most gentle bee colonies, I waited a few extra minutes for the bees to react to the smoke and then I opened the colony. Surprisingly, the bees behaved very well during the "bee show" and it turned out to be a very successful experience. However, I did not hear of any comments from my friends and acquaintances about the early morning honey bee show. I assumed that my friends do not get up that early and watch the news on TV and neither do I.

Lesson learned. I consider Joe, his camera lady, and myself to have been very fortunate that day. I would not recommend a beekeeper doing this under similar circumstances, because if conditions had been a little different, we could have been in for a tough morning show.

Carpenter Bees Don't Sting, or Do They?

Freshly graduated with a Master of Science Degree in Entomology, I figured I knew all the answers when it comes to questions on common insects like carpenter bees. When asked by my wife if carpenter bees sting, I assured her they would not sting humans. At that moment, we happened to be on the deck of our house where my wife was sunbathing. You might have guessed it, within five-minutes after saying that, a carpenter bee landed on the back of my neck and I quickly raised my hand and hit the bee and it stung me. I discovered that female carpenter bees do sting when provoked. However, their sting is rather mild similar to a honey bee sting, but the female carpenter bee's stinger is smooth and she can sting more than once, so be careful around those females.

Carpenter bees are about the same size as bumble bees, although the latter are extremely good pollinators and should be protected. Bumble bees are very hairy and have a structure on their hind legs called the corbicula which collects pollen. Honey bees have this same structure on their hind legs for the same purpose. This structure is not present on carpenter bees and they lack the hair on their abdomen, therefore their value as pollinators is less. Bumble bees are ground nesters whereas carpenter bees are wood nesters that bore holes, especially into soft wood like cedar and pine.

The carpenter bee is considered to be a pest around homes where they bore their nest entry holes and galleries, leaving behind their sawdust which can be a nuisance. Thank goodness, they are only active a few months in spring. The natural home of carpenter bees is in

Fig. 14 Female carpenter bee boring her gallery into the face of my front porch door

Fig. 15. Female carpenter bee preparing to enter her gallery entrance hole

dead wood found in the forest, so if you live in or near a wooded area where these bees are present, you can expect to see them around your house every spring.

Another form of exterior wood surface destruction is possible as a result of woodpeckers or squirrels searching for carpenter bee brood in nesting galleries. Complete destruction of boards is possible when this happens.

Fig. 16. Woodpecker damage to wood as a result of searching for carpenter bee brood. (Photo Courtesy of David Christopher)

Fig. 17. Closeup of photo above showing carpenter bees galleries ripped apart by a woodpecker searching for bee brood for food (Photo Courtesy of David Christopher)

Bumble bees have a hairy abdomen or rear end, whereas carpenter bees have a metallic or shiny black abdomen. Male carpenter bees have a white spot on their face and can be identified easily by this feature when in flight. The male has no stinger and cannot protect himself, but the males are very territorial and will pursue other males in the vicinity of their nest sites. They will also swoop down at humans who enter their territory, but they are harmless. Female carpenter bees are not territorial and will have a more direct flight pattern from their nest entry hole to and from foraging areas.

Carpenter bees are not truly solitary bees as two-three females may occupy the same nest site. Their entry hole will branch off with different females utilizing separate branches for laying eggs. Some females may overwinter for two seasons with older bees dying by early August.

The carpenter bee's main diet includes nectar and pollen, so they are considered   pollinators. But, they have mouthparts or mandibles that allow them to bore and make perforations externally on flower corolla tubes to feed on the nectaries thereby avoiding the normal pollinator process of coming in contact with plant pollen on flower anthers. This behavior is called "nectar robbing" and other pollinators like honey bees will follow suit and feed through the same holes made by the carpenter bee. This is a very serious problem to growers who produce blueberries and can result in lower crop yields due to poor pollination. Blueberry growers should attempt to eliminate carpenter bees on and near their farms.

Fig. 18. Male carpenter bee with characteristic white spot on front of his face.

Since carpenter bees are considered much less beneficial as pollinators and can damage wood siding or decking materials, man will often attempt to control them around structures. My favorite method of control is swatting them with a tennis racket, however this activity has little value toward long term control since I am killing only the males which are territorial and are easy picking for elimination. A more effective control method is spraying the entry holes with a pesticide labeled for carpenter bee control and plugging the holes the next day with steel wool, wood dowels or wood fillers like DAP® plastic wood. This method will provide long term control by killing the female carpenter bees and preventing the next generation from emerging. Another effective control method is the use of traps placed in areas frequented by carpenter bees looking for new nesting sites in early spring. There are many carpenter bee traps marketed in the US and some home-made trap design plans are available online.

Lessons learned. Only female carpenter bees can sting, but rarely do so unless provoked. Carpenter bees collect nectar and pollen from some trees and plants in spring, so they are known as being somewhat beneficial as pollinators. However, the lack of having hairy bodies as do honey bees and bumble bees makes the carpenter bee a less valued pollinator. Carpenter bees are better known for being a pest because of their destructive nesting habits around structures and for being known as "nectar robbers" around blueberry farms. The scientific name for the Eastern Carpenter Bee, which is found throughout the eastern US and Canada, is *Xylocopa virginica*.

Cicada Killers on Campus

On more than one occasion, I received calls from university employees about large bees that were harassing students and other people walking on campus. The large bees in question were cicada killers (.6-2.0 inches in length) which are in the wasp superfamily Vespoidea, and are known for having a "constricted waist" between their thorax and abdomen. Cicada killers are solitary and make their nests in the ground. They are fairly harmless, but the females do have a stinger, so I would not try to provoke and upset them in any manner.

I've often been amazed at the activities of the cicada killer. For a couple of years or so, we had some very active colonies on the Clemson University campus which received a lot of attention.

The cicada killer, sometimes referred to as the cicada hawk, emerges at about the same time of year (August) as its prey, the cicada which is its main source of food. The cicada killer normally catches its host in midair, attacking inflight and quickly stinging and immobilizing the adult cicada. Cicada killer adults feed also on plant nectar and plant sap and are active for only 60-75 days. Females are often twice as large as males.

On campus, the cicada killers would often make their nest in beds of liriope or monkey grass which was growing adjacent to walkways. The males are very territorial and are often seen fighting in midair with other males for dominance which creates quite a scene at the height of their activities. The males do not have a stinger making them harmless to

Fig. 19. Cicada Killer, Sphecius speciosus. Note the constricted waistline between the thorax and the abdomen.

Fig. 20. Cicada killer captures its prey, the cicada. The adult prey is as large as the cicada killer making for quite a difficult haul back to their nest entrance.

Fig. 21. Cicada killer ground nest entrance.

the public, however they will often fly at oncoming pedestrians in their zeal and haste to protect their territory. This often resulted in much excitement and even fear from students and university employees who happened to be walking in the protective zones of the male cicada killers. My recommendation to the university grounds crew was to not attempt to eliminate the ground nesting insects because they were harmless and their territorial activities would cease in a couple of weeks.

Cicada killers only have one generation per year as only the immatures successfully overwinter in the ground. Females provision their nest with dead cicadas and lay eggs on their prey, then close their nest with dirt. Single burrows have been known to have as many as 10 or more nest cells.

The only time that I have seen cicada killers become a major problem is on fine lawns or golf courses particularly around bunkers or sand traps. Their unsightly nests were mostly an annoyance for only a few weeks when the adults foraged for prey.

Do Honey Bees Forage at Night?

Of course, honey bees do not forage at night. If my memory serves me correctly, this question never came up during my teaching career, mainly because who would think of asking such a bizarre question? And, I had never read or heard of honey bees foraging at night. Honey bees are dependent on the sun's location for communicating direction to a food source and are even able to utilize ultra violet light on cloudy days for the same purpose. So, of course honey bee foraging is dependent on the sun, therefore bees do not forage at night.

Then one summer evening, Jack Lombard and I visited Arthur Maxie who was a beekeeper in rural Oconee County. Arthur and his good friends Slim Taylor and Dean Boggs were extracting sourwood honey in his backyard honey house. The honey house was not "bee tight" meaning that foraging honey bees could enter the house during daytime hours and harass the beekeepers while they extracted honey. Therefore, Arthur and his friends chose to wait till late evening before extracting their honey when honey bees are expected to all be safe at home.

As I observed Arthur, Slim and Dean extracting honey under minimum light conditions, my friend Jack walked off heading toward some honey bee colonies that Arthur had placed along a nearby pasture fence. Jack returned soon and asked me to follow him back to the colonies. He said, "You will not believe what you are about to see."

The time was about 9:30 PM and darkness had set in with the exception of a beautiful full moonlit night. As we approached the colonies, we observed honey bees exiting the hive and flying off across the open pasture. Some bees were also returning to the hive. This occurred during the first week of July at a time when sourwood trees are in full bloom in that area.

Fig. 22. Arthur Maxie on right with beekeeper friend Slim Taylor. The pasture in the background is where I witnessed honey bees foraging at night.

We remained at the site only a few minutes, so I do not know how long the bees continued to forage that evening. But, we had just observed a rare event when honey bees were foraging at night. I regret not observing the foraging bees longer to see how long this unusual activity continued. I have now thought about this occurrence and have come to one conclusion. The right conditions were needed for this to happen including: a full moon night, calm and warm weather conditions, and a strong nectar flow from nectar producing plants or trees such as sourwood trees which provide abundant nectar flows some years. I doubt the moon or the moonlight played a major role in the bees' navigation, rather the bees were dependent on the controversial theory that honey bees use odor left behind in the air of previous bee flights to navigate to and from their colony when foraging.

<u>Lesson learned</u>. I'm certain that it is a very rare event for honey bees to forage at night, but I learned from personal experience, given the right conditions, that it is possible. Those conditions include a strong nectar flow, warm weather, little or no wind, and a full moon night that played an unknown role. Since this occurrence, I have read of one other report where "Bees were observed by Rubick (pers comm) in Central America foraging on warm nights under full moon" as recorded by Southwick.[21] I assume this was David Rubick who studied Africanized honey bees in Central America at that time. Other bees have been reported to forage at night, but only other tropical bee species.[21]

Fig. 23. Sourwood Tree. (Photo Courtesy of Paul Boone, South Port, North Carolina7)

Do You Know a Beekeeper Who Plays Golf?

Chances are that you do not know a beekeeper who plays golf. During my career in support of the beekeeping industry, I have met or known thousands of beekeepers, but I have met only a few beekeepers, less than 10, who also played golf. So, what is going on here, as I have often pondered over this fact, that so few beekeepers play golf? I have concluded that there may be several reasons why beekeepers do not play golf.

One logical reason is that beekeepers simply do not have time to play golf and when they retire, it is too late to take up such a sport that requires much patience, finesse and coordination. But there may be other deep-seated reasons why they do not play golf. Beekeepers just do not have a desire to pay good money to hit and chase a little white ball for hours without any kind of return for their efforts, other than to get exercise. At the end of their tiring workday, the last thing they'd choose to do is play golf.

Another possible reason is that the type of person who is attracted to beekeeping is not attracted to playing golf. Both golf and beekeeping are outside activities, but golf requires about four consistent hours for a full round, whereas beekeeping is not as time conscious and can be accomplished as time permits at a more leisurely pace.

Although I do not have data to back up my theory that so few beekeepers play golf, I'd highly suspect that there is a much smaller percentage of beekeepers who also play golf compared to the percentage of the general population who play golf. By the way, I do play golf now that I am retired and I enjoy the game. I have more time to play golf and more patience to learn the game now than during my earlier working life. I cannot say that my golf game has improved much over the past few years, but I do enjoy the exercise and fellowship with my fellow golfers.

Fig. 24. This Beekeeper Plays Golf. (Photo courtesy of Rick Glover)

Missing Watermelons and Trailer Too!

One of my graduate students had a research project on seedless watermelon pollination and, in the process of his field work one summer, he grew about an acre of watermelons. The project resulted in an abundance of watermelons produced that year. At harvest time, he had to take a large sampling of the melons and weigh them, test them for sugar content, and take the diameter and length measurements of each fruit. There must have been about 200 watermelons tested from the field, but there was an abundance of excess melons produced from the project that were not processed for analysis.

This project was conducted at our Cherry Farm Research Station which is off campus and somewhat isolated from the public. My graduate student was not on an assistantship, so he had to work on the research project without pay that summer. He came to me and asked if he could take the excess melons to a local farmers market and sale them. Not having funds to pay the graduate student for his labor, I agreed that it sounded like a good idea to me too, but that he would have to promise me that this conversation never took place.

My graduate student loaded up about 100 of the excess melons onto a trailer on a Friday afternoon and he rolled the trailer with melons to the backside of the farm for storage till the next morning. He had assumed that no one would notice the melons and they would be safe there. To his surprise, he came in early the next morning to take his melons to the farmers market and someone had stolen the melons and the trailer too!

Fig. 25. Watermelons just harvested. (Clemson University Extension Program Photo)

My student reported the theft to the sheriff's office later that morning, but to no avail. Unfortunately, the melons had no markings, so tracking the fruit would be very difficult. A sheriff officer did report finding an excess of melons of the right size and color being sold at a local farmers market, but the vendor selling the melons claimed to not know the person who had sold him the melons. So, that turned out to be the end of that investigation and story. I figured someone who was familiar with the Cherry Farm operations, perhaps a passerby, witnessed the melon harvest taking place that afternoon and came back that night and stole the melon-loaded trailer.

Confining or Tethering Animals Near Bee Colonies Not a Good Idea

It is never a good idea to confine or tether animals near managed or feral bee colonies. Animals sometimes have a way of upsetting bees causing them to be subject to attack by a large number of bees. Even if the bees are known for being gentle in the past, they can sometimes become defensive when aggravated by animals such as dogs, goats, cows, or horses.

Fig. 26. Confining animals like this near a honey bee colony is not a good idea. (Florida Department of Plant Industry Photos)

I remember well a good example of this situation that occurred in coastal South Carolina several years ago. A homeowner had two dogs confined in a small kennel in their backyard and a neighbor who lived behind them had honey bee colonies in their backyard. There was a privacy fence a few feet high between the two properties.

On a hot summer day, honey bees were visiting the kennel to get water from the dogs' bowl. We will have to assume that one of the dogs began snapping at the bees which led to additional bees coming to the small kennel and the situation got out of control with both dogs being stung repeatedly and no way to escape. The dogs' owner heard the frantic dogs barking and ran out the back door and opened the kennel door to release the dogs and take them inside her house. During the process she received several stings, also.

One of the dogs did not survive the ordeal, as it must have reached the venom damage threshold effect on its body during the episode. The other dog was stung several times but survived. Both dogs were very healthy prior to the incident, according to the owner.

The dog owner sued the beekeeper for damages including the value of the deceased dog and the cost of veterinary care for the surviving dog. The dog owner claimed that the honey bees that stung their dogs came over the privacy fence from one of the beehives in their neighbor's backyard and stung the dogs.

The beekeeper claimed that there was no way to prove that the bees came from one of their colonies and that they may have come from a nearby feral colony. The beekeeper noted that the privacy fence would not have allowed a direct path from their colonies to the dog kennel.

This incident occurred when Africanized honey bees were entering the US in Texas and that story was all over the news channels. I do remember that the beekeepers honey bees were tested for African genes and the report came back negative. The case went before a court hearing but was settled before it went to a jury. I was subpoenaed for the case along with my counterpart at North Carolina State University. Although I was in town for the case, I did not have to appear, but my counterpart gave a deposition. Soon after his deposition, the case was settled out of court. From my understanding, the beekeepers' homeowner insurance covered the damages and medical expenses. The insurance company shortly thereafter cancelled their coverage.

Lessons learned. Animals should never be confined in small cages or tethered when there are honey bees or other stinging insects like yellowjackets, hornets, or fire ants nearby. Healthy, free-roaming animals can run and get away from stinging insects. Beekeepers should provide their bees a fresh source of water especially during warmer periods of the year to prevent them from visiting neighbor's animal watering bowls or swimming pools.

Bee City, USA

Bee City has been a favorite stop (1066 Holly Ridge Lane, Cottageville, South Carolina 29435) for tourists for several years, but it has made its greatest impact on the younger generation. Archie and Diane Biering founded Bee City in 1994. The small-scale city was built in the form of little stores, town hall, post office, barbershop, church, school, police station, and other small structures, each being home to a colony of honey bees. Although Bee City is located in a rural area of Colleton County along the Edisto River, it claimed to have a population of 1,260,000 residents. (honey bees)

Fig. 27. Bee City, USA, Population 1,260,000. (Photo Courtesy of Diane Biering)

You will find Bee City in the coastal area of our state, less than an hour drive west of Charleston. Through the years, Bee City has expanded to include other attractions such as a petting zoo, a honey bee products themed gift shop, café, nature trail, classroom and nature center.

On some of my visits to Bee City, I found groups of young school aged children learning about honey bees and their value in the classroom. Educational beekeeping visual aids, including an observation hive, were found around the inside perimeter of the classroom which were used in the instruction. A workshop was found next door which allowed the children to make beeswax craft items, such as candles and ornaments. Over the years, Bee

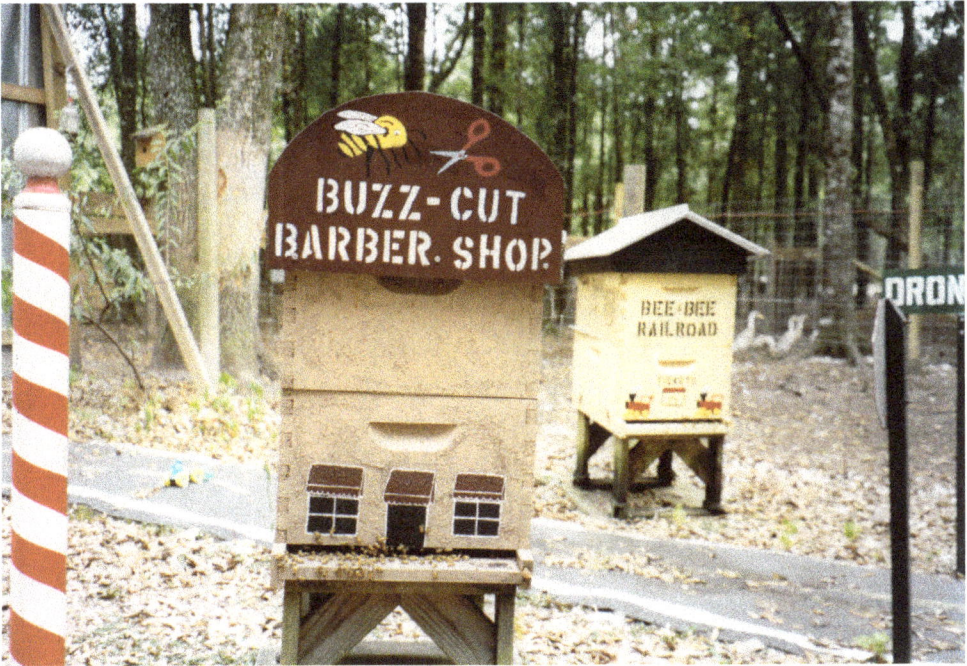

City became well known for school group visits as well as visits by boy scouts, girl scouts, church groups, and garden clubs. The classroom was also used for many years as a meeting room for the Lowcountry Beekeepers Association which was begun in 1990. Archie and Diane played a major role in founding the local association and Archie served as president for many years.

In 1994, Archie and Diane also managed about 100 honey bee colonies for honey production and pollination in the Cottageville area. They harvested the delicious tupelo honey and pollen and sold it in the gift shop. The gift shop featured other products such as creamed honey, beeswax candles, lotions, lip balms, caps and t-shirts.

Archie and Diane retired from Bee City in 2011, but their son (Scott) and daughter-n-law (Bridgette) are now owners of Bee City. They have increased the number of honey

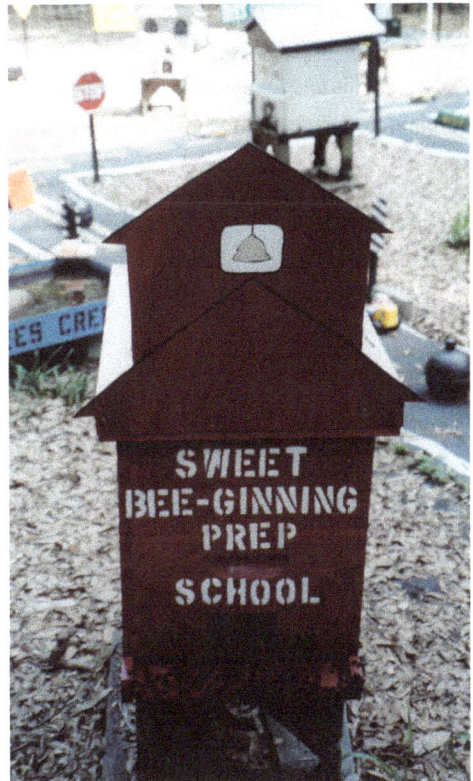

Fig. 28. Bee City School and Barbershop. (Photo Courtesy of Diane Biering)

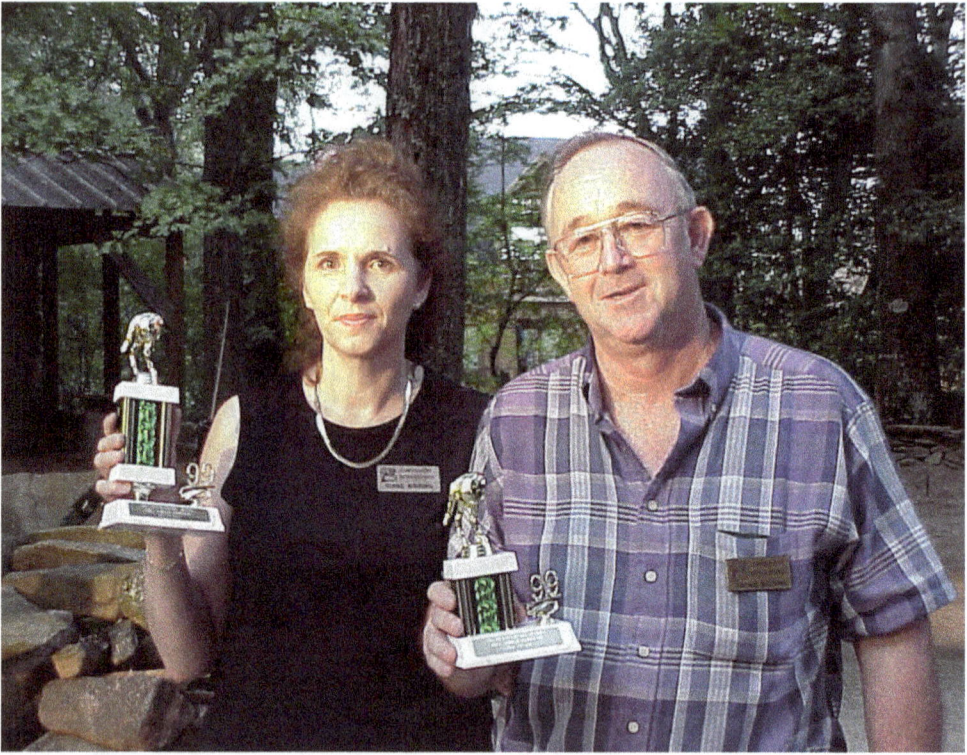

Fig. 29. Archie and Diane Biering, founders of Bee City, showing trophies won at the South Carolina Beekeepers Association Tall Tales Contest held at Clemson University in 1999.

bee colonies over the years and have added new animals, including many exotics. The name of Bee City has now changed to Bee City Honey Bee Farm and Petting Zoo. More than 40,000 school children visit this favorite attraction each year.

The Honey Bee Stinger

No book on "bees and man" would be complete without a story on the honey bee stinger. The honey bee has a great need to protect its colony from intruders, including man. The colony is a storehouse full of honey, pollen, and bee brood which must be protected at all costs for its survival.

The worker honey bee stinger provides the mechanism that is used to defend the colony. The stinger is barbed which becomes embedded in the intruder's tough skin. The stinger along with the seventh abdominal segment tears away from the bee's abdomen and remains behind in the victim. The last nerve ganglion and attached musculature remain behind pumping venom into the intruder's body. The sting shaft buries itself deeper and deeper into the victim as additional nerve endings may be hit resulting in the venom going deeper. The venom is the defensive agent made up of a complex mixture of peptides. The honey bee which has lost its stinger will soon die, but it is a small price to pay to ensure colony survival.

There is a very good chance that other bees from the same colony will be attracted to the same sting site to sting also. As the bee tears its body away from the stinger left behind, it releases a highly volatile chemical, isopentyl acetate, also known as alarm pheromone, which is stored and released by a membrane near the stinger. There is a good chance that a hive intruder will be stung more than once on the marked target site. A beekeeper can neutralize the alarm pheromone scent by smoking the sting site.

Colony intruders may include other honey bees from nearby alien colonies. Beekeepers use the term "robbing" when this event occurs which is normally a result of a stronger colony taking food resources from a weaker or stressed colony. The most serious threat to a honey bee colony comes from nearby stronger colonies which, during periods of dearth, seek to steal its honey stores. My good friend, Dr. Wyatt Mangum from Virginia, prefers to use the term "recycling" rather than robbing when this event happens. Regardless of the terminology, honey bees attempt to protect their colony stores by stinging intruders. However, they sting intruding bees in the soft body tissues that allows the stinger to remain intact without the barb becoming snared in the intruder's body. So, honey bees can sting other alien bees multiple times without losing their stinger.

It is true that adult bees from the same colony have the same distinct odor that is different than bees from other colonies. This trait allows guard bees at the entrance to quickly identify robber bees from other colonies. The common odor of bees in the same colony is not likely an inherited identifier, but it is acquired from the bees' environment within the colony. The cuticular waxes of the bee's exterior likely absorb the identifying odors, particularly from the food stores that have been collected from various floral sources. It is assumed that this identifying colony/bee odor can change over time as the foragers move onto other floral sources.

If you find yourself the victim of a stinging incident with a stinger left behind in your body, it is very important that you remove the stinger immediately, so as to minimize the amount of venom entering your body. It does not matter how you remove the stinger, just get the stinger out as quickly as possible. I'd suggest using a flat edge object such as a hive tool or credit card to scrape the stinger away within 15 seconds of the stinging incident. The honey bee is the only hymenopterous insect that leaves its stinger behind. Other bees including yellowjackets, wasps, hornets, and bumble bees have a smooth stinger which is not left behind in the victim. Therefore, multiple stings by the same bee are possible.

There are many sting site remedies recommended, but my favorite is common table salt mounded up on the sting site with a couple drops of water placed onto the salt. The moist salt removes the venom from the sting site by osmotic pressure. As with any sting remedy,

Fig. 30. Honey bee leaving its stinger behind in its victim

immediate application is extremely important. After a few minutes following the stinging incident, a topical remedy will be less effective because too much venom will have entered the victim. As for my recommended salt sting remedy, small salt packets from fast food restaurants are perfect for inclusion in a first aid kit.

In South America, where Africanized bees, often referred to as killer bees, escaped captivity in 1956 and began to be involved in major stinging incidents, bee scientists began research to select for honey bees with less propensity to sting. The researchers were successful in developing a strain of honey bees that were ineffective in their ability to sting an intruder. The stinger was modified to deflect when attempts were made to sting a victim. As the scientists were proudly reporting to a group of beekeepers their new strain of somewhat defenseless honey bees, they were met with some unexpected results.

The beekeepers were not in the least impressed with this new strain of honey bees because the bees would not be able to defend their colony against human thievery or robbing from alien bee colonies. So much for stingless honey bees.

Down through the ages, bees have been used by man for purposes other than honey production and pollination. The famous Roman poet Virgil, born in Italy in 70 BC, wrote much about bees. He shared the story that soldiers once tried to plunder his property, but most of his valuables were safe because he had stored them in and around his beehives.

Even today, beekeepers are known to hide their valuables in beehives which are known for being ingenious and burglarproof safes.[15]

The ingenuity of women as beekeepers is well illustrated in a story of an account of how Beyenburg in Prussia was named during the period of the Thirty Years War. Enemy soldiers were looting and ransacking the town, and stealing cattle, when they came upon a nunnery. The nuns seeing the advancing soldiers, hastily overturned their beehives that surrounded their homes and ran for cover. As the beehives lay dismantled on the ground, the furious bees began stinging everything in sight. The vandals soon retreated from the town as they discovered they had met a too formidable enemy.[3]

An interesting story was published in Bee World in 1933, when an old German beekeeper-farmer was held up by three thieves near his apiary. Since he was outnumbered, the man gave into the robbers and just at the opportune time, he struck a nearby beehive with his cane. The bees came out stinging everyone, but the beekeeper was less affected as he was accustomed to receiving stings. But, the thieves who were also stung many times, fled the scene to get away from the bees. The three men were apprehended down the street by police who easily identified them by the swollen sting marks on their bodies.[15]

Since the honey bee stinger produces a painful experience to the recipient, it should be no surprise that bees have been used in human warfare. This type of biological warfare is often referred to as a form of entomological warfare.[18] One of the earliest examples was throwing bees or their nests into caves to flush the hiding enemy into the open.[6]

In another example, a carved column found in Bucharest, the capital of Romania, shows the early Dacians defeating the Roman armored legions by throwing beehives down on them from the walls of their village.[14]

The advantages of enlisting bees and their relatives in man's military endeavors occurred during biblical times. In Exodus 23:28, reference is made to God's use of stinging insects to clear the Promised Land for the Israelites, "And I will send hornets before thee, which will drive out the Hivites, the Canaanites, and Hittites out of your way."[2]

The Naturalist's Library reported that in the 18th century that honey bees were used as a potent defensive weapon in the battle of Alba Graexa by Turkish forces. The Turks had succeeded in destroying portions of the wall when the foot soldiers were ordered to advance over the remains of the wall. As they began to breach the wall, they found that it was not defended by soldiers but by stinging bees. The Turk soldiers had not been trained to fight such an enemy and they were ill-equipped to confront such a force. The Turks retreated hastily from the wall that day.[15]

During the US Civil War, honey bees played a role in the Battle of Antietam on September 17, 1862. Attacking Federal troops, made up mostly of the rookie 132nd Pennsylvania, were advancing through a farmyard when they were routed, not by Confederate gunfire, but by stinging honey bees whose beehives had been shattered by rebel artillery.[11] The stinging bees proved to be more than most of the green soldiers could stand, as most of them "leaped and ran and slapped and swore" resulting in the regiment's confusion and disorganization. It took the combined efforts of the regimental officers, as well as General Nathan Kimball, the brigade commander, and his staff to get the boys out of the yard and back into their ranks.[15]

Another example of entomological warfare occurred during in WWI when honey bees were used to save the lives of Belgian soldiers who were barricaded inside a bee house. The attacking Germans were allowed to enter within a few feet of the house when the Belgians threw beehives full of bees at the Germans, sort of "animated hand grenades."[15] Not very long afterward, the Germans returned the stinging warfare as they defended themselves with honey bees against the British in the Battle of Tanga, which is known today as Tanzania in East Africa. The story goes that the German soldiers, as the Indian units along with the 1st Loyal North Lancashires approached, hid beehives in the dense vegetation and rigged the hive covers with wires. As the advancing British units unknowingly tripped over the booby trap wires, the covers of the hives pulled off and the bees promptly launched a maddening attack of their own. It was reported that the British soldiers were severely stung during their advance, but they did not retreat.[15]

There are more recent examples of honey bees used in warfare during the Vietnam War, also known as the 2nd Indochina War. Records indicate that both North and South Vietnamese units were guilty of using booby trap wires connected to Asian giant honey bee, *Apis dorsata*, colonies. The surprised enemy soldiers were met by the honey bees that are much more ferocious than our European honey bees. The larger bees attack in greater numbers, inflict more severe stings, and chase intruders further.[2]

According to the late Roger Morse (1927-2000), "Good judgment, a knowledge of bee behavior and the proper equipment make the much-feared sting a minor thing for the beekeeper. But it is interesting that this same good judgment and knowledge of bees which serves the interest of agriculture so well has also been used as a means of defense in many major wars."[15]

Lessons learned. The likelihood of getting stung by bees is increased if you are an out-doorsman. If you fall in the 1% of the human population that may have an anaphylactic reaction to insect venom, it is very important that you carry an Epi-Pen® with you at all times. Normal reaction to a bee sting by a person who is immune to insect venom includes pain, swelling and redness around the sting site. However, a person's immunity to venom can change over time, so it is a good idea to be familiar with the symptoms of anaphylaxis, so that correct action can be taken, if needed. An anaphylactic reaction to venom may include fainting, itching, rash, blood pressure drops suddenly, and airway narrows block-ing breathing. Other possible signs and symptoms include rapid or weak pulse, nausea, and vomiting.

Something Is Biting Me in My Bed

One summer, I had a young student work with me in a mentoring program sponsored by the university for up and coming high school scholars. It was very obvious this young man was extremely intelligent, but it was apparent that he had lived a sedentary city life style and had spent little time in outdoor activities in a rural environment.

I thought it was quite odd that he desired to work alongside a mentor who spent much of his time working with honey bees, especially in the summer which requires work outside in the brutal heat. Perhaps the young man did not know what he was getting into.

Anyway, he turned out to be a good worker, regardless of his background and did not hesitate to contribute in our research which was mostly fieldwork in the Clemson Forest.

However, after the first week of work, he came in Monday morning complaining that something was biting him in his bed. I thought this was very unusual since he was living in a university dorm room. He noted that the bites were very irritating and itched. The bites were mostly around his

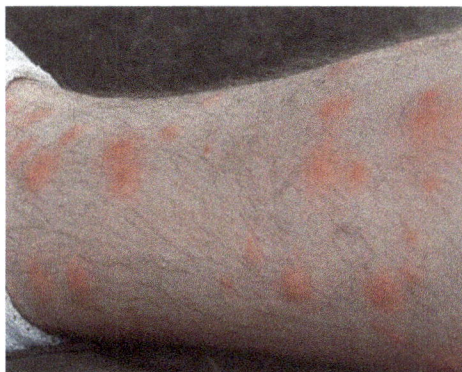

Fig. 31. Chigger bites can be very irritating for a week or two.

waist and on his legs. I had him role up his pant leg to show me what he was talking about. Apparently, this young man had never been outdoors in an area where chiggers are present. He had a pretty good case of chiggers on his body.

For the remaining few weeks that this student worked with us, we made sure that he used insect repellant containing deet, which is a good repellant for chiggers.

<u>Lesson learned</u>. This story reminds me of the annual chigger problem that we have often in our Southern forests, especially in stands of pine trees or stands of mixed pine and hardwoods. Chiggers, sometimes called red bugs, are in the mite family Trombiculidae. The larval stage of chiggers is the juvenile form which feeds on the human skin, lives there a few days, then drops to the ground to complete its lifecycle.

The feeding chigger site becomes very irritating in a day or two and is followed by intense itching and dermatitis. The irritation can last for a week or two. Normally, the problem cycles every year and the first chigger problems occur in late spring and can last through early fall. This is the time of year that we should spray our bodies, especially from our waist down, with a repellant before entering a stand of trees that we suspect may have chiggers. Chiggers can go undetected because they are less than 1/150th inch in length.

Yellowjackets and More

An Odd Hornet Nest

A photo of an odd hornet nest was forwarded to me by a homeowner because he had taken the rare photo and thought that it might be of interest to me. He was correct. The hornets had constructed the nest on the exterior side of his house under the soffit area of his roof. The nest went unnoticed for a while because it was located along a wall of his house that was rarely visited.

Fig. 32. Bald-faced hornet nest at base of soffit with distinct difference in color in lower portion of nest where European hornets had taken over nest construction. (Unknown Photographer)

What made this nest exceptional was the fact that the upper part of the nest was constructed by bald-faced hornets, *Dolichovespula maculata*, and the lower part of the nest was constructed by European hornets, *Vespa crabro*. Notice in the photo that there is a distinct difference in nest coloration with grey at the base or top of the nest and a lighter color of grey at the bottom. Apparently, the European hornets had taken over the bald-faced hornet nest in mid-construction and finished the job as photographed. The entire nest is made from paper-like material made from small pieces of dried leaves or bark. What also made this unusual was the fact that the normal nest site for European hornets is in the hollow of a tree a few feet off the ground and seldom on the side of a house.

The occasional observer might say that bald-faced hornets built the nest and abandoned it in winter and the nest was re-inhabited the next year by European hornets which finished construction of the nest. However, there are some issues with this train of thought. Bald-faced hornets do not overwinter as a colony in a nest. The abandoned nests make great conversation pieces when they are retrieved and hung in almost every park nature center and many homes of outdoorsmen. The nest must be retrieved fairly soon after the hornets abandon it or die, because predators, like squirrels or birds, often tear into them and consume the immatures that are left behind. Normally, a bald-faced hornet nest is made of pulp-like material and do not make it through winter in nature.

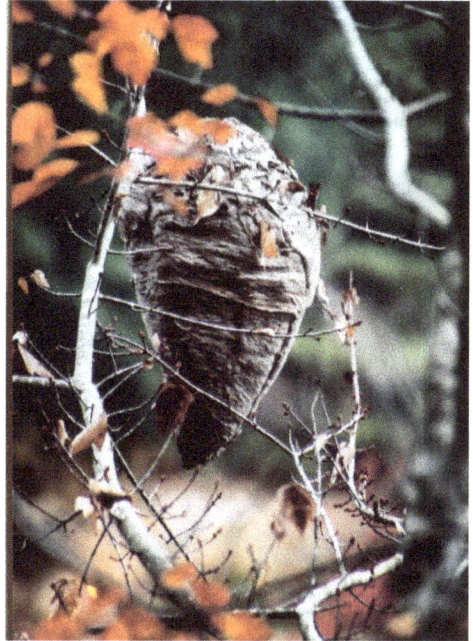

Fig. 33. Bald-faced hornet nest attached to a tree limb (Photo Courtesy of John R. Steedly, Naturecraft Photography).

Another possible explanation is that European hornets attacked the bald-faced hornets and over took the nest making it available for their colony needs. This idea also has some issues in that normally European hornets are the less aggressive insects. Another possible explanation is that the bald-faced hornets abandoned the nest or were killed by the homeowner and the European hornets re-colonized the nest and continued construction.

The bald-faced hornet is not a true hornet, but it is a black and white yellowjacket wasp that normally makes its hanging nest in a tree or shrub. However, nests can also be found on the sides of houses, sheds, or other man-made structures. The nest can measure up to 14 inches in diameter and 24 inches in length and may contain 100-600 workers. Bald-faced hornets can be dangerous to humans because when disturbed the females are quick to protect their nest and inflict a very painful sting. They also have a unique defensive mechanism in that they can squirt or spray venom from their stinger into the eyes of vertebrate nest invaders. The venom can result in temporary blindness to the intruder.

Bald-faced hornets are omnivorous feeding on flies, caterpillars, spiders, fruit, nectar, and other yellowjackets. They are also known in late season to forage at picnic shelters feeding on sandwich meats and sugary soda in trash cans. They have well adapted to heavy human population centers where their large nests attract a lot of attention from the public who often favor their removal. However, they are found to be beneficial when nesting away from human activities.

Fig. 34. Trophy-sized bald-faced hornet nest cut from a tree. (Photo courtesy: animaltrappingremoval.com)

Giant Yellowjacket Nest

A story of a giant yellowjacket nest, located in the low country of South Carolina, made headlines a year or so following hurricane Hugo which slammed into the state in September 1989. The hurricane resulted in some major damage to houses and outbuildings that happened to be in its pathway, especially in the coastal region of the state. The owners of one particular small shed had not opened the door of the building to inspect the contents for several months following the storm.

The owners were in for a big surprise when they finally got around to opening the shed door. A giant bee nest had been constructed from the ceiling and hung right down into the middle of the building. The owners backed up, closed the door carefully, and notified government officials of the giant nest they had discovered.

Shortly thereafter, an inspector came out to take a look at the nest and collected some bee specimens for identification. The bee sample was identified as yellowjackets. Yellowjackets are normally ground nesting bees, but in this case, they chose to build their home inside the shed which gave them protection from the outside elements. Normally yellowjackets do not overwinter as a colony, but only future queens successfully spend the cold months alone protected under a pile of brush or some other protective material. Next spring, the mated queens find a suitable nesting place and begin to raise the first brood.

There are exceptions in nature and this happened to be one. Apparently, this nest was occupied for more than one year due to its large size. I never heard of the outcome of this giant yellowjacket nest, but I'm sure the owners were very careful removing it from their shed.

Fig. 35. Giant yellowjacket nest (Unknown photographer)

Another Giant Yellowjacket Nest in the Upstate

I received a phone call from the Oconee County Agricultural Extension Agent one day to let me know they had discovered a giant bee nest that was built on the side of an earthen embankment along a rural road in their county. The agent assured me that it would be worth my while to pay a visit to the site because it was getting much attention by the public and the bees needed to be identified.

I drove to the location and sure enough it was a very large nest located on the side of the road making it very convenient for everyone to see. The bees were identified as yellow-jackets which was not unexpected because yellowjackets normally construct their nest in the ground. But what was puzzling about this nest was the tremendous size. If someone or an animal happened to disturb the nest intentionally or unintentionally, there was a good chance someone could get stung multiple times. An individual yellowjacket can sting multiple times because it has a smooth stinger and does not lose its stinger like honey bees that have a barbed stinger that is left behind and dies.

This was another instance where it appeared the colony must have been constructed over more than one season which is unnatural for yellowjackets in South Carolina. I never heard the outcome of this nest, but I suspect the landowner found a way to destroy the nest and its inhabitants likely during cold weather when the yellowjackets were inactive.

Fig. 36. Yellowjacket nest photographed from a safe distance.

Fig. 37. The same yellowjacket nest photographed from up close and from an unsafe distance.

The Honey Bee Hawk

The European hornet, *Vespa crabro*, sometimes mis-identified as a Japanese hornet, can be a problem in beeyards as a predator of honey bees. I have heard beekeepers refer to this predator as the "bee hawk" because of its feeding activities around honey bee colonies. The European hornet is the largest eusocial wasp native to Europe and is the only true hornet (genus *Vespa*) found in North America. It was imported into the US by settlers in the 1800s.

Fig. 38. Enlarged view of the European Hornet, Vespa crabro, known by some beekeepers as the "bee hawk."

The European hornet can often be observed foraging for honey bees in flight in an apiary as the bees are returning to their beehive. Upon close examination, the observer will note that only returning bees are captured by the hornet because they will normally be loaded down with rewards of nectar or pollen. After the hornet captures a single bee in flight, it will fly and light on a nearby tree limb, gnaw the appendages off its prey, and then chew and consume the remaining contents of the bee. It will repeat this process many times during the day. No doubt, the European hornet can play a role in reducing the number of adult bees in a colony.

In other settings, the European hornet can be beneficial as its diet consists of other problem insects, so no control is recommended especially if the nest is located away from humans. But to the beekeeper, the hornet is a pest to be managed if possible, especially if 3-4 hornets are consuming foraging bees in a single apiary. A typical colony can have up to 400 workers at their peak, which occurs in late summer or early fall. The hornet will normally be nesting in the hollow of a nearby tree sometimes in close proximity to the apiary. If the beekeeper can locate the nest, he should exterminate the hornets with a product registered for hornet control. I have also seen hornet traps made from a two-liter soda bottle, baited with grape jelly and sugar water. The soda bottle with the cap on and a quarter-sized hole cut in the shoulder of the trap should be placed near the beeyard to attract and kill the hornets because they do not escape the trap.

Relentless Yellowjackets

A beekeeper friend shared the following humorous story about an incident that he had with some yellowjackets.

He was cutting grass in his backyard with a push mower one afternoon and received a painful sting on his hand. As he looked down, he realized that he had disturbed an underground nest of yellowjackets that were swarming out of the ground. He ran immediately to a nearby small shelter, pulling his mower along with a couple of the yellowjackets flying close behind. One of the yellowjackets seemed to have gone up his britches leg, so he frantically beat the seat of his pants with his hand to kill the yellowjacket. However, he received another stinging sensation on his buttocks, so he made a run for his house.

As he ran to the back door and through his house, he yelled to his wife that he was being stung repeatedly by yellowjackets in the seat of his pants. He ran out the front door into the front yard where his wife helped him unfasten his bibs on his overall pants to remove them and to get rid of the yellowjackets that had been stinging him repeatedly all along the way.

By now, he noticed that his wife seemed to think this ordeal was humorous, so he asked her what was so funny about him getting stung by yellowjackets? She reminded him that she had fastened his torn back pocket of his overalls with a safety pin that morning and that the safety pin had come un-snapped during his ordeal and the sharp point of the pin was sticking him repeatedly as he beat on the seat of his pants to get rid of the supposed yellowjackets. The relentless-stinging yellowjacket sensation turned out to be the sharp point of the safety pin sticking him on his rear-end every time he hit his pants.

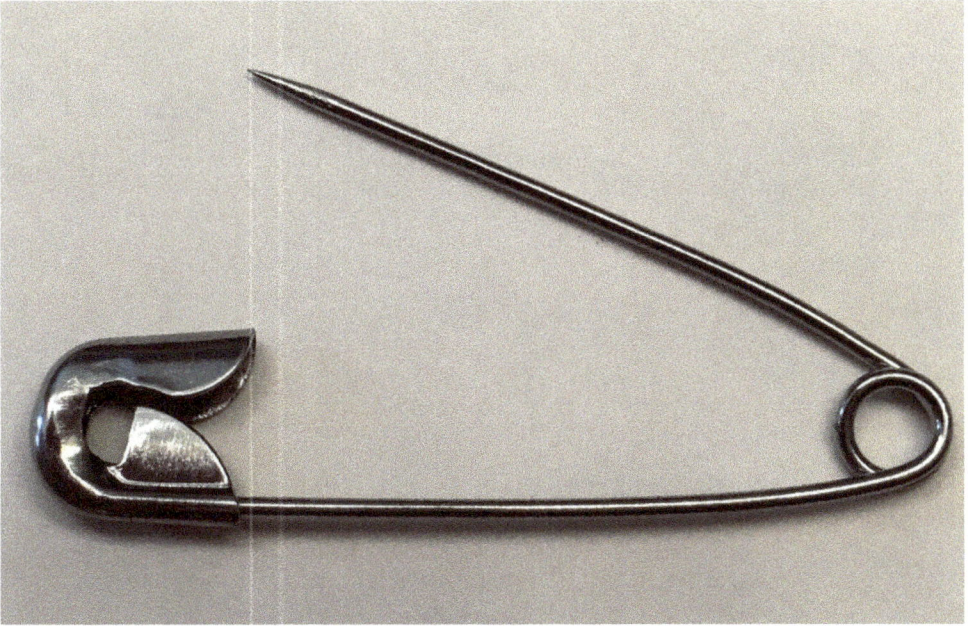

Fig. 39. Safety pin which was thought to be a yellowjacket. Ouch!

Honey Bee Swarms

The Myth About How to Settle a Swarm of Honey Bees

I have found that the average beekeeper does not have a good understanding of how bees hear and to what degree they hear. Bottom line, honey bees have a very poor sense of hearing compared to humans as noted in a follow up story later in this book.

When a colony of honey bees swarm, there will be much activity in and around the beehive. The swarm will emerge by forcing the queen to exit the hive entrance with most of the other bees. Prior to their exit, the worker bees are known to perform what is called "buzzing runs" by bees running along the comb in straight lines while vibrating their partially opened wings. The buzzing runs increase in intensity which results in much excitement prior to and during swarm emergence.

A few beekeepers have mentioned to me that they were able to cause a swarm of dispersed bees, as shown in the photo below, to settle upon hearing a metal to metal clanging, better known as "tanging." When honey bees swarm, they move as a group of bees dispersed in the air similar to a cloud moving through the sky. Scout worker bees which have found a new nest site lead the swarm in a directed effort normally in a straight line to the new nest site. The scout bees release their Nasonov gland pheromone to guide the airborne swarm. The queen is somewhere in the swarm and gives off a pheromone (9-oxo-trans-2-decenoic acid) to maintain cohesion in the air that helps coordinate and guide the flying bees to their destination.

Fig. 40. Swarm of honey bees in flight (Photo Courtesy of the Loudoun Beekeepers Association, Virginia)

The myth that has been passed down from beekeeper to beekeeper is that if you bang together two metal objects such a pan or pot and a spoon, the bees in flight will hear the noise and settle nearby, thereby providing the beekeeper the opportunity of capturing the swarm of bees. If truth be known, those beekeepers who claimed success in settling a swarm by making the clanging noise, would have been just as successful without the clanging because bees are largely insensitive to airborne sound of this nature.

Fig. 41. Illustration of the myth showing a man beating (tanging) two metal objects together in an attempt to settle a swarm of honey bees.

A swarm of honey bees just emerging from a hive will normally settle into a cluster on a nearby tree limb or bush within 50-150 feet of the parent colony with or without tanging. The clustered swarm of bees will remain there for several minutes or even hours. On rare occasions, they may remain there for months if the scout bees are unable to locate a new nest site. But normally within an hour or so, the bees will disperse into the air and travel to the new nest location.

The true story of this metal to metal clanging idea around a bee swarm was used centuries ago by beekeepers as they chased or followed a dispersed swarm of bees in flight. When a beekeeper discovered a flying swarm of bees, he ran with the swarm while clanging two metal objects to alert other beekeepers of his ownership of the swarm. Sometimes a honey bee swarm traveled for quite some distance across various owner properties before settling and this was simply a method of proving bee swarm ownership and protected the beekeeper from trespassing. This swarm chasing technique originated in certain archaic laws.

As an alternative method of settling a swarm of bees, a procedure was reported in an early version of the book, ABC and XYZ of Bee Culture.[19] A swarm of bees in flight can be made to cluster sooner than normal often by spraying fine particles of water on the bees from a hand forced pump. The fine particles of water may simulate rain or may dampen the bee's wings impeding their flight or both, resulting in a very deciding effect of forcing the bees to settle earlier than normal. The hand forced pump of water may also be used when retrieving a swarm that has settled beyond the safe reach of the beekeeper. The pump sprayer nozzle should be adjusted to apply a stream of water directly on the cluster making the bees dislodge and take wing and finally settle on a lower limb.[19]

Honey Bee Swarm Arrives by Express Air Mail

The Boy Scouts of America had a merit badge in beekeeping at the time this story occurred in about 1995. I was contacted by a local family who knew me quite well. Their young son, Christopher, was a member of the local boy scout troop and they asked if I'd be interested in working with him in attaining the merit badge in beekeeping. So, of course, I agreed to guide the young man through the process of earning his badge.

I met with Christopher and we discussed all the requirements for attaining this beekeeping badge. One hands-on requirement was to catch a swarm of honey bees and place them into a beehive thereby starting a new honey bee colony. Or, as an alternative way of meeting this requirement was to start a new colony of bees by dividing an existing colony and making what we call a split. In order to make a split, the beekeeper takes half the frames with clinging bees from an existing colony making sure not to take the queen from the parent colony and introducing an additional queen to the queenless split into another beehive, thus making two colonies from one.

Christopher and I agreed that we would much prefer the option of catching a swarm of bees to start up a new colony. A swarm of honey bees is normally made up of the old queen and about 60% of the bees that leave and fly off from a parent colony in a well-defined group. The swarm will usually settle in a nearby tree or suitable temporary support as a cluster and may remain there a few hours to a few days depending on how long the scout bees take in finding an acceptable home. The parent colony from which it swarmed will have raised a few queens and they emerge about a day or so following the swarm's exodus. Other smaller swarms may exit the same colony soon taking with them one of the virgin queens. If no other swarms emerge from the parent colony, the worker bees allow the first queen to emerge and she will often sting and kill rival queens in their cells before they emerge or she will fight it out with them if they successfully emerge, so that only one queen will head up the colony.

This was during the month of May which is a good time of year to catch a swarm of bees because the months of April and May are the two main months for honey bees to swarm in South Carolina. I told Christopher that we should get the word out that we needed a swarm of honey bees and that we would gladly come and capture a swarm on a moment's notice.

About two weeks later, Christopher called me all excited and said that there was a swarm of honey bees hanging underneath his grandmother's mailbox. She lived next door to him and I'm sure the mailman was a little surprised to see the bee swarm that contained probably 10,000 bees hanging in a cluster underneath the mailbox.

Fig. 42. Honey bee swarm on mailbox (Photo Courtesy of City of Chesapeake, VA)

I hurried and loaded a hive body with frames into my truck and headed over to his grandmother's house. Sometimes a honey bee swarm will move onto a more permanent home, so I knew that I needed to hurry. Sure enough, I drove up and there the bees were hanging from the bottom of the mailbox. Christopher and I placed the open hive body on the ground below the swarm, shook the mailbox, and the bees fell straight down into the box. That evening we moved the bee colony to Christopher's backyard next door where he was able to manage them that summer.

Lesson learned. Now what are the chances of this happening? We needed a swarm of honey bees and they came special express from an unknown bee colony in the area and landed of all places, next door on his grandmother's mailbox. That is a pretty amazing story that I have often pondered many times over as to how all that could have happened at just the right time and just the right place. WOW!

The Daycare Center Swarm

My first spring on the job at Clemson University I received a call from a nearby daycare center. A highly excited lady explained to me that a swarm of thousands of honey bees had settled in a small water oak tree in their playground. I told her that I would load up my equipment and should be there in about 15 minutes. This was my first swarm retrieval job, so I made sure of taking along every conceivable thing that I might need.

I drove up to the daycare center parking lot and sure enough there was a huge swarm of honey bees hanging in a cluster about eight feet off the ground. I could not believe my eyes because some the bees were flying around the tree which was located in the center of the playground. The children were playing in the playground and no one had been stung. However, I knew that honey bees when swarming are very docile and are not prone to become defensive, particularly when they have just swarmed from their parent colony.

Fig. 43. Large Swarm of Honey Bees in Tree. (Photo Courtesy of South Carolina Commercial Beekeeper Eric Mills)

I opened the playground gate and walked over to the lady who looked to be "in charge" and I told her that she would have to take the children inside while I captured the swarm of bees. She said that would be fine, but could I please do her a big favor and allow the older children, the five-year-old's, to sit quietly along the fence and watch me capture the swarm of honey bees.

I reluctantly agreed to allow the 15 or so older children to sit quietly and watch. I thought, WOW, what a teachable moment. Well, what is an extension bee specialist better to do than demonstrate and provide a new and exciting experience to a group of children? At that moment, it did not occur to me that these children would watch every move that I made and would remember these moments for the rest of their lives.

So, I went to work putting on my protective bee suit coveralls and gloves, just to be sure. I even lit my smoker to help calm the bees, if needed, especially if something went wrong with my plan. Remember this was my first time to catch a swarm of bees. I placed my hive body (box) right below the swarm, just like the book said. I carefully climbed onto a folding chair and reached to grasp the limb the bees were hanging. I gave the limb a good shake and all the bees came tumbling down into the box below.

Since this occurrence, I have captured many swarms of bees, but this was the largest swarm of bees that I can remember capturing during my entire career. There must have been 30,000 bees in the swarm and when they fell to the box, there were just as many bees on the ground than fell into the box. I stepped down from the chair and started scooping up bees with my hands, placing them in the box, which I later discovered was unnecessary. Because, if the queen is in the box, the bees on the ground will march straight into the box entrance eventually like soldiers, having sensed the Nosonov pheromone given off by the other worker bees.

To my extreme surprise, I felt a few bees crawling up my legs and I dared not make any quick moves because there was a real good chance the bees might start stinging me in a very sensitive area of my body. In my haste to get my protective equipment on, I had failed to secure my pant legs inside my boots. BIG MISTAKE! The problem was that there were so many bees crawling on the ground in all the chaos. I have since learned that when honey bees come to a vertical surface they tend to climb upward. My legs provided many of the disoriented bees a vertical surface to climb, to my dismay.

But the interesting part of this story is to remember that I had 15 five-year-old children watching my every move from about 20 yards away. If these bees start stinging me, there will be movements of my body that I had not planned. So, I gently walked over toward the teacher and let her know the show was over and that she should now take the children inside, while I finished the job. She agreed and took the children inside the building and I had assumed out of sight.

After the children had gone inside, I had to remove my coveralls and let down my other pair of pants to pick the bees off my legs one bee at a time. After re-securing my pants, I finished the job and felt pretty satisfied about capturing a very large swarm of bees for a good addition to my growing university apiaries. Most of the bees were now in the box or heading in that direction, apparently the queen fell into the box and stayed there which is a good thing to happen. If the queen had not stayed in the box, she would have likely flown and settled back in the tree and all the bees would have followed her.

However, I learned that my assumption about the children being out of sight was wrong when I went back early the next morning to pick up the box with all the bees inside. I knocked on the daycare center front door and was invited to enter. I noticed a couple of the ladies there were kind of laughing when I entered. It was then and there that I noticed a large picture window on the playground side of the building. This was where the children had gathered to watch the rest of my swarm capture show outside.

The significance of this story rang true when a month later one of my good friends from another department came over to talk to me as I was refueling my truck at the university motor pool. I could tell immediately he had something humorous to tell me. He said his five-year-old son came home from day school a few weeks ago and told him an interesting story about the "beeman" who took the bees away at school that day. He claimed that I had become his little boy's hero and from that day on he would not wear anything but "Fruit of the Loom" underwear!

Lessons learned. When bees are involved, you never know what to expect, so it is best to be prepared and enjoy the moment when life throws you a curve. For new beekeepers, I'll have to admit that I had on far too much protection that day. The extra pair of pants was unnecessary and normally you do not need a smoker when capturing a swarm. Gloves are optional because normally swarms of honey bees are very docile as they have no nest to defend. However, I have heard some horror stories from beekeepers capturing swarms that had been hanging on a tree for several days. In that case, the bees were not able to find a proper nesting site and have transitioned to making the location their home and they will defend it.

I have found European honey bee colonies nesting exposed to the elements for long periods of time, but it is very rare. I witnessed a huge colony of honey bees nesting high in a tree in the coastal town of Denmark, South Carolina. The colony remained there for at least three years. I collected an adult bee sample from the colony and mailed the bees to the USDA Beltsville, Maryland Bee Lab, where they were confirmed as European honey bees.

Stories About Beekeepers

The Beekeeper Who Took My Smoker

A beekeeper located in northern Greenville County, South Carolina called me at my office one day early in my career. He had talked to a Greenville County Agricultural Extension Agent and discovered that I was available to inspect honey bee colonies, particularly if the beekeeper suspected that he had a disease or pest problem.

He claimed that someone had given him several honey bee colonies recently and that he would like for me to pay him a visit and inspect his bees for disease problems. He noted that he did not have much experience as a beekeeper, but that he was anxious to learn more on how to manage his bees. He had little advice on the history of these bee colonies and had not opened any of the beehives which were located in the middle of a woodlot on his property. I agreed to pay him a visit.

Arriving at his home in the early afternoon at the appointed time, I drove up in his yard and he met me and joined me as we drove toward his beeyard. I sensed a strong alcohol odor as he sat down in the passenger seat of my pickup. Perhaps he thought that having a drink of alcohol would help to calm his nerves.

We drove up to his beeyard and I parked the truck about 25 yards from his bee colonies. I suited up and lit my bee smoker, which is used to help keep the bees calm. I noticed that he did not have a bee suit, but he was heavily dressed with an overcoat for such a warm afternoon and he wore a very flimsy homemade bee veil with ball cap.

I opened up his first bee colony and checked it for any problems and found none. He stood behind me and observed what I was doing, but he did not offer to help. As I opened the second colony, I noticed the bees became very defensive immediately and appeared to be out of control, so I reached for my smoker to calm the bees and it was not where I had placed it. Apparently, several of the angry bees had entered the man's flimsy bee veil and he had run to my truck with my smoker in hand. I have found over the years that people can do unexpected things when chased by stinging bees.

The beekeeper had left me to fend for myself without my smoker which is a necessity when working hostile honey bees. But, I quickly managed to reassemble the colony and I walked to my pickup truck and found my beekeeper friend had climbed into my truck cab, closed the door and windows, and had my smoker inside his veil trying to pacify a couple of angry honey bees. By this time the beekeeper had a few bee stings on his head and he was not a very happy person. I drove him back to his house, making sure that he was feeling alright before I departed. I have a sneaky feeling that he passed those bee colonies onto another beekeeper.

Fig. 44. Beekeeper's best friend, the bee smoker.

Lessons learned: Always be prepared when working with honey bee colonies, particularly if you are not familiar with the temperament of the colonies or if you have little experience as a beekeeper. And, remember to always have a veil on and have your smoker lit and ready for action whether you think you will need it or not.

The correct use of smoke will tend to lessen the aggressiveness of a more defensive colony. The exposure of bees to smoke will result in many of the bees inside the colony engorging themselves with honey from the comb, making them less prone to defend the colony. Other benefits of smoke around bees is that it masks the beekeeper's alien scent and it masks the alarm pheromone released by the guard bees making their alarm system less effective.

One question that is often asked by novice beekeepers is, "What is the best fuel for a bee smoker?" My favorite fuel for a smoker is frayed dry pine needles, sometimes called pine straw, which you will find on paved roads, especially in the fall of the year. You will often find them in the median or the side of a road where car tire treads have smashed the needles, producing maximum surface area. The frayed pine needles found in roads are often extremely dry which is another bonus making lighting a smoker much easier. Pine needles can be stored for long periods of time when placed in an onion bag or similar bag in a dry storage place. My favorite storage container was a 5 gal. plastic bucket with lid which I carried in the bed of my work truck.

Fig. 45. Excellent smoker fuel: frayed pine needles found in the median of a paved road in October. Note: Beekeepers are advised to be extremely careful of oncoming traffic when collecting pine straw on busy roads.

Wagon Ride With Arthur Maxie

My good beekeeper friend the late Arthur Maxie invited my family and me to his farm one spring day for a visit. So, my wife Kathy, daughter Lauren (8), daughter Lindsey (7), son Jordan (5), and I accepted his invitation and decided to take a picnic lunch along for the visit.

Arthur was about 75 years of age at that time and lived a simple life in rural Oconee County, South Carolina in the Wolf Stake community where he owned a farm of about 40 acres. Arthur never married and lived with his mother till she had passed a few years earlier. Arthur never had a driver license, so he depended on others for transportation. Although, he did have a tractor which he drove on his farm where he had a sizeable garden that he sold vegetables from to the public and he grew and sold sweet potato slits in the spring. He also managed about 50 honey bee colonies which he highly valued. His main source of income was his social security check along with some income from selling garden produce and honey to the public.

Fig. 46. Arthur Maxie (on right) selling a jar of honey to a customer.

Arthur was located in an area where his bees could make sourwood honey, which by the way is known by many honey connoisseurs as the most delicious honey produced in the United States. One of the favorite stories he would tell was when he was a young lad, he and his buddies would cut down "bee trees" and extricate the bees when the weather was too rainy to work on the farm. A "bee tree" is a tree which a colony of honey bees has set up residence in a hollow of the tree. By the way, this is the natural home site of honey bees in a feral setting. After cutting the tree down, Arthur and his buddies would cut the tree apart and carefully take out the bees and place them into a beehive. This activity along with catching swarms were his two major sources of increasing the number of honey bee colonies in his operation.

The day of our visit to Mr. Maxie's farm came, so Kathy and I loaded up our three children and headed out to Oconee County. When we drove up to his house, Arthur greeted us and told us to come on board the farm wagon that he had hooked up to the back of his tractor. By then, our children were super excited sitting on the bales of hay in his wagon.

As we proceeded to ride down behind his house, we passed a rickety old barn which had a fenced-in area where Arthur housed his chickens. One separate area was for holding his roosters. Arthur had informed me during an earlier visit that those were his prized roosters that he entered in "cock fights" on Saturday nights, which was an illegal activity at that time. For those of you who do not know what this pastime involves, two rival roosters each fitted with metal spurs are released into an arena and most of the time they fight till one dies. Arthur was well known in his area for having some roosters which had "ruled the roost" you might say in the arena. I did not bother to explain to my children why Mr. Maxie had so many roosters that were kept in their individual cages.

We next passed by Arthur's garden on our way to the beautiful meadow where he kept several of his honey bee colonies. This is where we stopped for our picnic. It was a glorious day with several varieties of flowers and trees in bloom. The bees were very active coming and going from the nearby beehives as we were in the middle of the major nectar flows of the year.

We finished up our picnic and headed back to Arthur's home in his wagon. I think Arthur enjoyed our visit as much as we did. My children truly enjoyed our visit to Mr. Maxie's farm and they can still remember this special experience.

Fig. 47. Arthur Maxie standing beside a row of Anise Hyssop, a good plant for bees.

Jack Lombard: A Mountain of a Beekeeper

The late Jack Lombard (1927-2014) was a true mountain beekeeper who lived in Mountain Rest, South Carolina in northern Oconee County. Jack retired after 33 years of service at the Walhalla National Fish Hatchery which was located a few miles north of his home. Jack who lived into his late 80's before passing was also known for his regular Saturday night square dance calling at the nearby Oconee State Park. He was simply a local legend.

Jack's home place was a favorite stopping point for many who traveled from Walhalla up SC Hwy 107 into the Blue Ridge Mountains. He proudly claimed of being born in the same house that he now lived. Jack had a large fenced in area where he kept deer and wild turkeys which he had mostly raised from abandoned fawns or young captured turkeys. The fence was about 6 feet high which the deer and turkeys could have escaped easily, but why escape into the wild when conditions were so good inside the fence, according to Jack.

Fig. 48. Jack Lombard, Oconee County mountain beekeeper, on his 4-wheeler.

My first introduction to Jack, who was a well-known Oconee County beekeeper, came early in my career when I stopped by for a short visit to introduce myself. I knocked on the front door and his wife came out and told me he was out back of the house making apple cider. I walked into his backyard and found him in a small wooden shed that he had setup to crush apples and make apple cider. Jack had an apple orchard on his farm where he harvested fresh apples in fall and he would make apple cider to sale to the public. That day he gave me a gallon of his freshly made apple cider which turned out to be delicious. I knew then that I had picked a good day to meet and to visit with Jack.

Jack always welcomed my university undergraduate beekeeping course students for a visit to hear experiences of "the life of a mountain beekeeper." One time we drove up in his driveway and exited our vans and did not immediately see him. However, we were very surprised when Jack came speeding over the hill riding his 4-wheeler. You just don't expect an 80-year-old man to be riding a 4-wheeler so fast, but the students loved it.

Jack typically discussed mountain beekeeping, especially the production of his favorite honey, sourwood, which is produced only in higher elevations. He talked about the challenge of protecting his honey bees and their honey from black bears, which roamed that area of our state. He showed the students how he protected his 10-15 honey bee colonies that he had strategically placed onto a flatbed trailer outside his bedroom window. He had a strand of wire secured around the trailer with a cowbell attached. If a bear approached during the night and made a move for the beehives, the cowbell alarm would sound and Jack would ease his bedroom window open. He never said what his next step was to protect the bees.

After talking to the students more about beekeeping in the mountains, Jack would lead the students over to some of his wild animals such as wild hogs and raccoons, but the student's favorite stop was the rattle snakes which Jack had captured in the wild around his farm and surrounding mountains. He kept the snakes in a circular metal walled container that you could look over the side and observe the snakes which dinned in holes in the ground. To say the least, the snakes were the hit of the day for just about every class that I took to visit Jack.

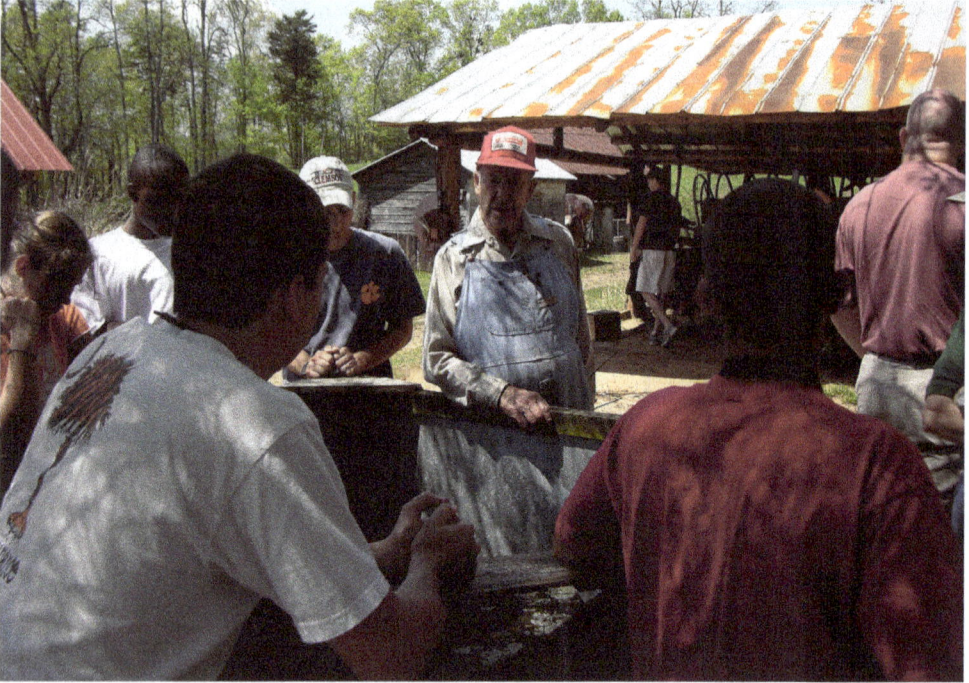

Fig. 49. Jack entertaining my students at his rattlesnake pit.

Fig. 50. Jack's rattlesnake inside his snake pit.

Jack and I became good friends over the years and he invited me to bring some of my university honey bee colonies to his farm the first day or so in July which is when the sourwood nectar flow begins in his area and lasts for about 3-4 weeks. I would load up about 10-12 colonies on my trailer and head to Jack's home place. He allowed me to place my trailer inside his deer fence to keep them safe from bear activity. Some years, I made a good crop of sourwood honey which I gladly shared with Jack. Some of my university and non-university friends were spoiled on the mountain sourwood honey which I gave them on special occasions.

Paul Brown's Famous Honey Bee Colony

The late Paul Brown (1920-2018) lived just across the northern South Carolina state line in North Carolina near the Charlotte, NC Airport. I met Paul when he had attained senior age status, however he was very active as a beekeeper in both North and South Carolina at the time. Paul was a lead instructor for the York County (South Carolina) Beekeepers Association which hosted an annual beginner level short course for several years. Paul was always ready to advise and help beekeepers. He was a very dedicated beekeeper and instructor who supported the beekeeping industry in both states. In fact, in the year 1997, Paul received the annual "Beekeeper of the Year Award" from both the North Carolina State and South Carolina State Associations. This prestigious award from both states will likely never go to the same beekeeper again.

Paul was well known for his beekeeping skills and was proud of his record making honey bee colony that produced 490 lbs. of honey in a single year in Mecklenburg County, North Carolina. That is the equivalent to 180 quarts of honey from a single honey bee colony. That was quite an achievement, given that Paul was a stationary beekeeper who did not move his bee colonies to take advantage of nectar flows in other regions during the year. He claimed to have harvested 8,000 lbs. of honey from 40 colonies for a 200 lbs. average in 1999. Paul was an inspiration to many beekeepers throughout the Carolinas.

Fig. 51. Paul Brown and his record-breaking honey bee colony

Paul also introduced me later in life to the idea of raised garden beds at his retirement home in Hendersonville, North Carolina. One of the primary benefits of raised garden beds is that you do not have to bend over as much when working in your garden. Paul took great pride in producing garden vegetables which he donated to the kitchen in the retirement village where he lived. I returned home from one of my trips to visit with Paul and built myself a raised garden bed in my backyard. I discovered that deer liked my raised garden bed too. I photographed one feasting on my vegetables one night, so I had to place some anti- predator netting around the garden which worked fine in keeping the deer out.

Huck Babcock: A Well-Known Commercial Beekeeper

The late Havilah Babcock Jr. (1920-2010), known to most as Huck, lived in Cayce, South Carolina, which is located in the midlands of our state. Huck was a free-spirited commercial beekeeper who at one time had the largest number of bee colonies in the state. He sold package bees and queens throughout the eastern US for many years. Huck was the owner and front-man of Blue Ridge Apiaries and he had 2-3 workers who did most of the physical labor, especially one beekeeper named Frank Brown. Huck had 30-40 production apiaries or beeyard locations within a 75-mile radius of Cayce, particularly in the Congaree Swamp area, located just east of Columbia. He had other apiaries located in Lexington County, Richland County, Lee County, Calhoun County and Orangeburg County. At one time during his beekeeping career, Huck moved some of his bees to Florida and Virginia for various nectar flows to make honey.

Fig. 52. Huck Babcock, owner of Blue Ridge Apiaries, Cayce, South Carolina.

Fig. 53. Frank Brown, employee, Blue Ridge Apiaries, Cayce, South Carolina.

Huck sold many package bees to beekeepers north of the Mason-Dixon Line. He told a story of one New York beekeeper who bought packages of bees from him for several years in a row and the beekeeper always called and complained about his order every year. Huck decided he had heard enough from this complaining beekeeper, so he told Frank to go out back of their warehouse and make up a package of bees from a colony that they had avoided for years. Apparently, this colony had some bad genes for excess defensiveness and no one bothered to go near it. Frank suited up heavily and was able to find the queen and included it into the package of bees which was shipped that day to the New York beekeeper. Huck claimed that he never heard from this beekeeper again.

Another story that Huck told at many beekeeper gatherings was a story that occurred when honey bee tracheal mites were first discovered in the US. At that time, Huck was fond of moving a few hundred of his bee colonies to Florida to harvest some of their delicious orange blossom honey. Unfortunately, the Florida Dept. of Agriculture bee regulatory agency quarantined the state's beekeeping industry and no bee colonies were allowed to enter or leave the state's borders.

According to Huck, this regulatory action did not deter him from moving his bees to Florida. He and his employees loaded up a big enclosed trailer with bee colonies and placed boxes of apples at the very back of the trailer, so that an inspector would only see apples upon inspection and would allow them to pass through the checkpoint station. Mr. Spann Leitner was one of Huck's employees at the time and he drove the truck loaded with apples and the unsuspecting bee colonies.

Huck's luck ran out at the first Florida agricultural check station. The inspectors must have sensed something was up with this load of apples. Perhaps, a few escaped bees flew out of the trailer during the inspection. The inspectors asked Spann if there was something other than apples in the trailer. To Huck's dismay, Spann was likely one of the most honest persons in the state of South Carolina. So, when asked, Spann said there were honey bee colonies in the front of the trailer.

Huck had to make an unscheduled trip to Florida as he was responsible for the illegal contraband. According to Huck, he was called before a local judge near the Florida border where he was able to give the judge a few hundred dollars cash and he and Spann drove back to South Carolina with the load of apples and bees. Huck claimed that Spann was a good employee, but that he was just too honest.

Huck loved to make sourwood honey in the mountains of Oconee County, so annually he would move 200-300 colonies there and leave them for the month of July. Bears became an increasing problem for beekeepers who moved their colonies to northern Oconee County which is native habitat for black bears. One measure of bear protection that Huck used once was to place a battery-energized radio into an empty beehive box and set the radio on high volume to a 24-hour station. Huck was convinced the loud music would deter bears from coming close to his beeyard. I informed Huck that he was going to have bears dancing to the music in his bee yards while they tore through his colonies eating bee brood and honey. As a backup, he installed electric fences around the bee yard to protect them from bears just in case the battery failed to energize the radio.

Fig. 54. Huck Babcock fine tuning his radio inside a beehive box to protect the bees from black bears. Note: double bear proofing with radio noise and fence.

For some years, unfavorable weather or other conditions prevented a good sourwood nectar flow, but Huck claimed that any honey that passed by the Oconee County Courthouse in Walhalla on his way going home to Cayce was sourwood honey. This claim got Huck in hot water a few times because some people favored sourwood honey and they knew if the honey was produced from another nectar source or blended with other honey. For payment to the mountain region landowners' use-rights of their property, Huck would give each of them a case or two of honey annually at the end of the year. On one of those low sourwood honey production years, Huck made the mistake of giving Jack Lombard, a mountain beekeeper, a case of non-sourwood honey without explanation, for use rights to his land. Jack was not happy over receiving some non-sourwood honey that year, so Huck was never able to place his colonies on Jack's property again.

Fig. 55. Sourwood honey production beeyard in Oconee County prior to bears becoming a problem in the area. In other words, no fences were required to protect them. These 40 Babcock colonies had just been unloaded.

Charlie Holden: The Honey Bee Charmer

An upstate South Carolina beekeeper, the late Charlie Holden (1940-2014), was well known for being an expert remover of honey bee colonies from structures. He prided himself in doing honey bee colony removals without the use of any personal protection other than his bee smoker. I have seen Charlie on the job taking a few stings with no effect, even pulling a few stingers from his bald head.

Fig. 56. Charlie (on right) and helper removing a honey bee colony from an exterior wall of a house.

Charlie had a background in building construction with particular expertise in electrical work. This experience helped when it came time to remove an unwanted honey bee colony because the wall of the structure often had to be dismantled to get the bees out and he had to follow up with repair of the wall to its original condition before the job was complete.

NOTE. *A honey bee colony located in the side of a structure results from a swarm of bees that have settled and set up housekeeping in the side of a house or other building. Honey bees favor selecting a structure that has a minimum of several square feet of void inside the wall. The bees normally will select a nest site on the eastern or southern facing side of a structure because it offers the colony warmth due to more sun exposure in winter. Another discriminator is selection of a nest site that has a small entrance that can be protected from predators and other enemies of the honey bees.*

Charlie had an entertaining little technique of killing persistent and annoying individual honey bees that would not leave him alone during colony removals. He raised both hands as though preparing to clap his hands, but rather he would agitate both hands quickly which drew the bee to the activity and when in Charlie's sight, he would close his hands quickly with a hard-clapping motion catching the bee in a deadly clasp without being stung.

Fig. 57. Charlie doing his "deadly charm" on an aggravating honey bee.

Over the years I received many calls particularly from homeowners asking if I removed bees from structures. After explaining to the homeowner that I could not take bees from their structures for several reasons, one being the liability issue, I would give them Charlie's contact information. Charlie charged for his services at a minimum of $500 for a basic bee colony removal job. However, Charlie shared with me the fact that he had removed many bee colonies over the years without charge to homeowners who simply could not afford to pay. Thanks Charlie for being a good beekeeping ambassador.

Spann Leitner: "Baby Doll"

The late Henry Spann Leitner (1909-2006) lived in Winnsboro, Fairfield County, South Carolina and worked in the beekeeping industry for most of his lifetime. When I called his house phone, his wife would answer the phone and call out the name "baby doll" for Spann to come to the phone. I met Spann toward the end of his life and learned to truly appreciate his dedication and love of honey bees and the South Carolina beekeeping industry.

Spann was instrumental in organizing the first state beekeeper's association in South Carolina which is known now as the South Carolina Beekeepers Association. He served as the first vice president of the association and hosted the first meeting at his home in Winnsboro on March 22, 1975.

On one of my visits to spend a little time with Spann, he called my attention to the fact that his bees had paid for his house by way of income that he derived from his honey sales. He said, "Do you see my truck? The bees paid for it too." He said, "Do you see my dog over there? He will not ride in my truck with me anymore after we had an accident going down a nearby dirt road to visit a beeyard a few weeks ago."

Spann was known to many other local bee-keepers as one who would sell his honey to the public too cheap. He told me that he charged a fair price for his honey and did not care what other beekeepers thought about his asking price of honey. He had

Fig. 58. Spann Leitner or "Baby Doll"

many local honey customers who would bring their empty honey jars to him and he would fill the jars with fresh honey right from his holding tank while they waited. He charged $.50 per pound for the honey which at that time was about half the going price for whole-sale honey. Apparently, Spann had been rewarded well in life from his chosen profession, regardless of the price of his honey.

Fig. 59. Spann Leitner in one of his home apiaries.

One other rule that Spann shared with me was that he was happy with harvesting only one super of honey per year from each colony which is about 40-50 lbs. Spann claimed that the remaining honey a bee colony produced was for their own needs.

Lesson learned. No doubt, Spann Leitner will be remembered by many as one of South Carolina's finest beekeepers. He pioneered in organizing the first state beekeeper's association in the state and assisted many new beekeepers in learning how to keep honey bees and how to enjoy working with them.

Fig. 60. Spann Leitner, "South Carolina Beekeeper of the Year 1989".

Ever Heard of Burt's Bees?

The late Burt Shavitz (1935-2015), the co-founder of Burt's Bees, is probably one of the better-known beekeepers in the US. His fame did not come from his great work as a beekeeper, but from the way in which he was portrayed to the public through his honey bee product line. Burt's birth name was Ingram Berg Shavitz which he changed unofficially to Burt following his high school graduation. He was born to a Jewish family in Manhattan and grew up in Great Neck, New York, served in the US Army[20], and moved back to Manhattan where he became a photographer for Time and Life magazines.[12]

In the 1970s, Burt came to the realization that he was terrified of growing old in a dingy apartment in the big city, so he decided to move to a rural setting near Dover-Foxcraft,

Maine. At this time in his life, he decided to learn the art of beekeeping, to live an easy lifestyle, to grow a beard, and to grow his hair long.[12]

Burt began his beekeeping career selling his honey by the gallon off the bed of his Datsun pickup truck parked along the side of the road. He marked on his beehives the words "Burt's Bees" to lessen the chance of them being stolen. Then, in 1984, he met a hitchhiker, Roxanne Quimby a mother of twins.[22] Their relationship began as a couple and not as a business partnership. Roxanne was an out-of-work waitress with a fine arts background.[9] Burt claimed that from time to time Roxanne needed a man, and that he was happy to meet that need. According to Roxanne, her new friend lived an idyllic lifestyle in the wilderness, including making his home inside a small turkey coop.[22]

Burt soon introduced Roxanne to a book which was filled with beeswax recipes for making candles. Eventually, Roxanne partnered with him to expand their product line to add beeswax candles, made from beeswax leftover from his honey processing. She sold their beeswax candles for $3 a pair at a local craft fair. The first year their sales grew from $200 at a single craft fair early in the year to $20,000 by the end of the year and Roxanne had bigger ideas.[22]

Their first product production facility was an abandoned schoolhouse which they rented from a close friend for $150 per year. In 1989, their sales skyrocketed when their candle line was added to a New York City hip boutique named Zona.[9] This led to the addition of more employees to meet demand and Roxanne jumped at the chance to add a line of homemade personal care products. They incorporated the company in 1991, the same year that Burt's Bees Lip Balm became its best seller with Burt owning one-third of the company and Roxanne owning the remaining two-thirds.[22]

However, Burt loved the simple life and was content with the meager sales, but Roxanne wished to grow the company which she did. Burt's major contribution to the development of the company was providing himself as the face of the company. As Burt's Bees became more successful, Burt and Roxanne's relationship and vision became

Fig. 61. Burt's Bees "Beeswax Lip Balm" Became the Best Seller in 1991.

Fig. 62. Recent Marketing Package: Burt's Bees "Lip Balm" Multi-Pack in a Sam's Club Store.

more strained.[12] According to a documentary published in 2014,[20] Burt noted that "Roxanne wanted money and power, and I was just a pillar on the way to that success."

Fig. 63. Burt's face on products was his major contribution to the company. Burt with wild hair and a bushy beard looked very much the part of an eccentric beekeeper, according to an article on the company's history.[9]

Burt was still active in the company up until 1993 when annual revenues topped $3 million. They, or at least Roxanne, decided it was time to move the company south to a state with lower taxes and more potential workers to keep the business growing.[22] Potential sites were considered in New Hampshire, North Carolina, Florida, and Tennessee with North Carolina being chosen because of its aggressive lobbying efforts.[9]

In 1994, Roxanne decided to move the company headquarters to Durham, North Carolina and Burt decided he had enough of the business life, so he volunteered to leave the company, according to one report. But, another report claimed that Burt was forced to leave the company after having an affair with a young employee. Regardless, during this time, the couple's relationship became even more strained and Burt decided it was time to move back to a simpler life in Maine.[22]

An 18,000 square foot production facility was opened in Creedmoor, North Carolina in late 1994. The first retail store was opened in Chapel Hill, North Carolina and by 1998 Burt's Bees product line was sold in 4,000 locations with sales exceeding $8 million. The product line began to show up in stores like Bath and Body Works, Target, and Cracker Barrel. By the year 2000, the company's revenues were over $23 million[1] and the company relocated yet again to a 105,000 sq. ft. building at Durham's Keystone Office Park near the Research Triangle.[9]

In 1999, Roxanne made a good investment by buying out Burt's one-third of the company share by purchasing him a house in Maine for $130,000. Burt sold the house in Maine within a few months likely because he missed living in his turkey coop which he converted and enlarged to about 12 feet by 20 feet.[22]

Burt's Bees annual revenues soared from 2000 to 2007, from $23 million to $164 million which resulted in Burt losing out on a huge future payday. In 2004, Roxanne reportedly sold 80% of Burt's Bees to an investment company, AEA, for $173 million. Clorox purchased the company for $925 million in 2007. Roxanne's share of the Clorox deal was $300 million, as reported by the Associated Press.[20] According to Roxanne's family, they passed along $4 million to Burt and claimed that everyone involved in the company was rewarded fairly from the sale of the company. Occasionally through the years, Burt would re-enter the business world briefly and promote Burt's Bee brand at special events for Clorox. Reports

Fig. 64. Burt's Bees Multi-Product Pack.

indicated that the company paid Burt an undisclosed amount of money each year for using his name and image on its product line. By 2008, Burt's Bees employed 380 employees.[22]

Shavitz died of respiratory complications in July 2015 at the age of 80. Although his life was filled with strife, Burt's journey through life was filled with joy in his sunset years as he lived in Parkman, Maine, with his three golden retrievers and had little need for electricity or running water in his enlarged house which was used once as a turkey coop. According to Burt, "The magic of living life for me is, and always has been, the magic of living on the land, not in the magic of money."[22]

Burt liked to reflect on his simple life that he had everything that he needed including 37 acres of land, fields and woods that he could watch hawks and pine martens. According to some who knew Burt well, he was happiest when alone on his land in Maine, which was far away from company executives, consumers, and money.[12]

According to a 2014 documentary "Burt was a hippie making a living by selling honey when his life was altered by a chance encounter with a hitchhiking Roxanne Quimby." According to Burt, "I had no desire to be an upward-mobile-rising yuppie with a trophy wife, a trophy house, and a trophy car, I wasn't looking for any of those things. I already had what I wanted. No one has ever accused me of being ambitious, he joked."[20]

Fred Deer: A North Carolina Beekeeper

The late Fred Deer of Cary, North Carolina was a good friend of mine and senior–aged beekeeper at the time of this story. Fred was a very knowledgeable beekeeper and had many years of beekeeping experience. You did not have to be around Fred very long to learn of his extreme competitive nature. He loved competition and I learned that he did not like to lose, regardless of the circumstances.

Our two state's beekeeper associations began meeting jointly every two years at one of their spring meetings and this tradition has continued through the writing of this book. The North Carolina State Beekeepers Association hosted the first joint meeting in Charlotte, North Carolina in 1991 and the South Carolina Beekeepers Association hosted the second joint meeting in Florence, South Carolina in 1993. These joint meetings turned into great opportunities for our beekeepers to come together for good fellowship and great learning opportunities because we could afford to invite some well-known national speakers to such a large gathering of beekeepers.

One favorite event at these earlier joint meetings was the Carolina Bee Bowl which pitted four members of each state association in a competition for their beekeeping knowledge. I learned very quickly that Fred Deer was the driving force for the North Carolina team. It was apparent that the North Carolina competition had been ongoing for years prior to our joint meetings and his beekeepers were well trained for such competition.

Fred was always busy getting all the competition equipment setup and tested which included setting up a table on each side of the stage. At each table, four positions were setup with a push button and light bulb for each participant. When a question was asked, each of the eight panelists had the opportunity of being the first to attempt to correctly answer the question after being the first one to depress their button causing their light to flash. Points were given to the team whose beekeeper was first to correctly answer each question. These Bee Bowl competitions were very spirited with good attendance and much excitement.

If I remember correctly, the North Carolina State Association team won the Bee Bowl competition the first three joint meetings hands-down. Fred Deer seemed to be the coach of the North Carolina team every year and he took great pride in their winning the competitions.

The fourth joint meeting was held in Myrtle Beach, South Carolina and was hosted by the South Carolina Beekeepers Association. My counterpart at North Carolina State University, the late Dr. John Ambrose (1944-2015), served as the master of ceremonies for the competitions and he and I would come up with equal number of questions in advance for all the Bee Bowls.

I had the feeling that Dr. Ambrose was determined to have the South Carolina team win this year's competition since we were hosting the joint-meeting and South Carolina had never won a Bee Bowl competition. Dr. Ambrose requested early on that I come up with some specific questions which would require someone from South Carolina to know the correct answers. By this time, I highly suspected John had something up his sleeve.

The night of the competition arrived and the score was running fairly close during the Bee Bowl and then Dr. Ambrose started asking those really specific South Carolina questions which tilted the game and the South Carolina team edged out to win their first Carolina Bee Bowl with great excitement from the South Carolina side. This seemed to be a highly spirited competition this year and the North Carolina coach Fred Deer was not happy over the outcome. Fred was well aware that Dr. Ambrose had tilted the game in favor of South Carolina.

After the competition that night, I saw Fred in serious conversation with Dr. Ambrose and even the next day Fred was still fuming over the outcome of the Bee Bowl. He just would not let this event pass without some choice comments. Fred let me know that he would single-handedly like to challenge the South Carolina Bee Bowl team to a rematch at the next joint meeting.

To my knowledge, this was the last joint meeting to have a Bee Bowl competition. So, Fred never got a chance to redeem himself and his North Carolina State Team.

Steve Taber III: Renowned USDA Honey Bee Scientist

The late Steve Taber III (1924-2008) grew up on the campus of the University of South Carolina, Columbia, SC, where his father was Head and Professor of the Department of Geology. Steve showed a keen interest in honey bees early in life as a young boy when he sold drone honey bee pupae as fish bait to local fishermen. As a teenager, Steve worked during the summer months with commercial beekeepers in New York and Wisconsin where he learned many of his basic beekeeping skills and developed an appreciation for honey bees. For his first beekeeping experience in 1941, Steve worked for Mr. Elton Lane in upstate New York where he worked six, ten-hour days and his pay was $30 per month.[23]

Steve graduated from University High School in Columbia, SC in 1942 and joined the Navy a few months after graduation. He served 3 years in the Navy and received an honorable discharge in 1945. Following his time in the military, Steve attended the University of Wisconsin, Madison, and studied under C.L. Farrar. Steve later worked for USDA as an assistant to O. Mackensen at the USDA Honey Bee Lab, Baton Rouge, Louisiana.[23] During this time, Steve pioneered in the development of artificial insemination and queen breeding for the purpose of developing disease resistance and gentle bee colonies. He became an authority on the subject and authored a book later in life titled "Breeding Super Bees." After working at the bee lab in Baton Rouge for 15 years, Steve was transferred to the USDA Bee Research Center in Tucson, Arizona where he later retired.

After retiring from USDA, Steve moved to Vocaville, California, where he and Tom Parisian cofounded "Taber Honey Bee Genetics"[23] in 1978. He later moved to southern France where he lived several years and continued his queen breeding research. While in France, Steve savored the international prestige as a queen breeder and traveled throughout Europe giving presentations at many beekeeper conventions. He enjoyed living in France, but moved back to his home state of South Carolina and lived in the small town of Elgin, which is near Columbia. There he lived a simple life and would attend our South Carolina Beekeepers Association summer meetings in Clemson. Steve gave a few presentations at our state meetings. He was always fond of presentations that focused on the subject of queens and one year he helped teach a short course on queen rearing, which was well attended and received.

Fig. 65. Steve Taber III on right with Dr. H. Shimanuki, USDA Bee Scientist, in center and his wife Susan on his left at a South Carolina Beekeepers Association summer meeting held at Clemson University.

Steve's articles and publications continue to be referenced by honey bee scientists in many parts of the world. He published his research in many journals and magazines including the Journal of Economic Entomology, Journal of Apicultural Research, Beekeepers Quarterly, American Bee Journal, and Bee Culture.

Steve was a free-spirited bee scientist who will no doubt be remembered for his outstanding research contributions to apiculture. Many students who worked along beside Steve over his long career are now leaders in the world of beekeeping research today.

Steve once shared with me a story about an experience he had during his later years while working at the USDA Bee Research Center in Tucson, Arizona. Richard Nixon served as President of the United States at that time and Steve let it be known that he did not care for President Nixon and always thought of him as a crook. All Federal Building Headquarters including the Tucson Bee Research Center had a photo of the serving President posted in the foyer. Steve had a habit of tilting President Nixon's photo frame to one side every morning on his way to his office. The lead administrator of the bee lab found out who the guilty party was who was messing with the President's photo. He called Steve into his office and told him to keep his hands off the photo or there would be consequences.

Fig. 66. Steve Taber III.

According to the story, this did not deter Steve. He remembered his long-time good friend Dr. Murray Blum who worked at that time at the University of Georgia Department of Entomology where he was well known as a pioneer in the study of chemical communication by insects. Murray told Steve long ago that if he ever needed a powerful foul-smelling odor that he had just the chemical. So, Steve called Murray and told him to send a vial of the smelly chemical that he had mentioned years ago because he had a special application for it in Tucson. Steve received the small vial in the mail and proceeded to open the vial lid and saturate a cotton ball with the smelly chemical and carefully attached it to the backside of President Nixon's photo early one morning. However, as he was rehanging the photo to the wall, the glass-framed photo fell to the floor and shattered into a hundred pieces with the chemical spilling on the floor. Steve did not have the time or patience to clean up the mess. The terrible odor spread to all parts of the building which had to be evacuated for an entire day to neutralize the foul odor and clean up the mess.

The lead administrator called Steve into his office the day after the incident and there were consequences, such as Steve not being able to travel at USDA expense and he retired a few years later, according to what Steve shared with me.

Beekeepers Can Be a Strange Lot Sometimes

Case 1. During my career working with the beekeeping industry, I found that beekeepers can be a strange lot at times. For instance, after presenting my normal 30-45-minute powerpoint presentation to a group of beekeepers, I would follow with a question and answer session, expecting to get questions on the topic that I had just covered. However, I could expect a beekeeper normally sitting in the back of the class to raise his hand and ask some off the wall basic beekeeping question that was totally unrelated to the impressive presentation that I had just delivered. All that I could do was smile and try to answer his question and move on. I learned early on in my teaching career that most beekeepers are particularly prone to learn more from hands-on training than from sitting back and listening to my presentations.

Case 2. Our state beekeeper's association had an executive committee made up of all officers of the association and representatives from many local associations. The offices of the state association were President, Vice President, Past President, Secretary and Treasurer. We had a somewhat contentious discussion at one of our committee meetings which really upset the gentleman who held the office of Past President. After our meeting, he informed me that he would like to resign his office of Past President. My question to him and it still is: how does one resign as Past President of an organization? I've never heard of someone being replaced as Past President of an organization. Just think, how would you feel if someone asked you to accept the office of Past President of an association?

Case 3. One night at a local beekeeper's association meeting, I gave a presentation on a new product, named Bee Quick®. I thought this new product was a great replacement for Bee Go®, which was an impressive product that clears honey bees quickly from their honey stores, but it has a terrible odor that is very offensive. The new product had an odor that smelled like an almond extract which was much more pleasant to use to remove bees from their honey supers. At the end of my presentation, an older beekeeper stood up in the back of the room and said "Dr. Hood, I will not be using this new product because if I did my wife would not be able to find me in the summer, because I use Bee Go® and she is accustomed to me smelling like the disgusting foul odor of this product."

Case 4. Beekeepers have this thing about trucks, especially pickup trucks. Most beekeepers do have a pickup truck to haul their equipment around and, in many cases, it doesn't matter what the truck looks like, as long as it is dependable. The appearance of the truck is not that important and paint is optional as long as it is not rusted out and does not have a roof leak. In my travels throughout the state to speak at beekeeper's meetings held in unfamiliar places sometimes, I could always spot the building where we would meet by the overabundance of pickup trucks in the parking lot.

Fig. 67. Beekeeper L.W. Rabon of Horry County and his beekeeping truck.

Case 5. Bee beards are common when large groups of beekeepers come together. I am always amazed when someone attaches to their bare face thousands of honey bees each having a stinger. Why they do this is beyond my imagination. I have heard that they use very young bees that show little propensity to sting and the bees are well fed, engorged with food. A caged queen is attached to the beekeeper's chin and the bees are poured onto a removable platform that helps create the beard of clinging bees. Cotton balls are placed in the beekeeper's nostrils and ears, however I've heard that a few stings are expected normally during the episode. I guess the showmanship of doing this is worth the cost of receiving a few stings.

Fig. 68. Laurence Cutts of Chipley, Florida. (Photo courtesy of Jeffrey Lotz, Florida Department of Agriculture)

Another Photo of Laurence Cutts, who just can't get enough bees (Photo courtesy of Laurence).

Case 6. One day, one of my helpers and I were moving a trailer full of honey bee colonies from the mountains back to Clemson when we came upon a driver's license check station setup by two highway patrolmen. As we moved slowly forward behind other vehicles to be checked, I noticed that several of our bees had escaped their beehives and were flying around our trailer. As we approached the patrolman, he took one quick look at our trailer of beehives and the few bees hoovering over the trailer, and he immediately waived us past the check station without wait or inspection. Apparently, this highway patrolman took the initiative and made an exception to his license check.

Case 7. There are few walks in life that you will find people of so many diverse backgrounds and educational levels involved in the same pursuit as beekeeping. I have found doctors, nurses, lawyers, ministers, school teachers, common laborers, school dropouts, as well as prison inmates who enjoy working with honey bees. This makes for an interesting group when they all come together to share their perspectives and ideas. Quite often, beekeeping is passed down from generation to generation. However, since the mid-1980s, there have been so many new challenges come along such as new pests and disease. Much of the basics in beekeeping have not changed, but the new challenges can make beekeeping tough especially to the novice beekeeper who is depending on his previous generation for their expertise needed to be successful. However, I do not see a shortage of new beekeepers learning how to keep honey bees and becoming involved in this fascinating pursuit from all walks of life.

Case 8. I have found that new beekeepers can sometimes come down with a bad case of what I call "bee fever." No doubt, the study of honey bees and beekeeping can be fascinating and exciting, especially to the beginner. The new beekeeper may begin to spend an inordinate amount of their time in pursuit of this new avocation of beekeeping and neglect other important responsibilities. My advice to novice beekeepers is to keep a good perspective on all of life's responsibilities, and make sure they do not get caught up spending too much of their time in this pleasurable and rewarding new adventure.

Fig. 69. Young men discovering the fascinating and exciting world of honey bees. (Photo courtesy of Bob Bellinger, Clemson University)

Case 9. Smoking is allowed when beekeepers come together for a "smoker lighting contest." The smoker is a metal cannister with bellows attached which is used to calm the honey bees as a beekeeper opens a beehive. For the competition, a smoldering fire is built inside the fire chamber and the bellows are pumped to blow smoke out the upper opening of the smoker. The object of the smoker lighting contest is to have beekeepers compete to see who can create the most smoke with their smokers in a given amount of time. The competing beekeepers alight their smokers at the same time and are allowed to pump the bellows for a brief period of time after which they sit their smokers on the ground and then wait another few minutes. A judge will then select the two smokers which continue to provide the most smoke and selects a winner and runner-up. Normally, these contests create a tremendous amount of smoke as the cheering audience tries not to be downwind of all the smoke.

Fig. 70. Scene from a smoker lighting contest in progress.

Case 10. I had talked to a senior-aged lady by phone a few times about some basic questions she had on beekeeping. Apparently, this lady had been given several honey bee colonies and this was her first year of beekeeping. She was in the process of requeening some of them and she needed much advice. One day she called me up and said that she had received in the mail some new queens in their individual shipping cages. She had become fascinated with the caged queens as they sounded as though they were fussing at each other. She had placed the caged queens on her dining room table and had spent hours listening to the noisy queens, claiming that it was the most exciting time that she had experienced in a long time. She informed me that I needed to engage in the past time also. I never did hear back from the lady as to how successful she was in beekeeping. At this time in my career, I was extremely busy with various projects and I failed to fully appreciate what this lady was experiencing. In the past, I had heard new queens making these sounds while holding them in their individual cages in preparation for introducing them into colonies, but I had never wanted to do this activity for hours. The queen fussing noise that she heard was what is called "piping." It is a high note sound like "tee-tee-tee-tee" that a queen makes by vibrating her wing muscles.

For the Beekeeper

Anxiety Over Cutting Up a Fallen Bee Tree

This story occurred about a year after Hurricane Hugo ripped through coastal South Carolina on September 21-22, 1989. The category 4 hurricane made landfall just north of Charleston with winds of 140 mph with gusts up to 150 mph. Widespread destruction resulted in the path of this major storm with very high winds which were even felt in the upstate where I live.

An excited senior-aged hobby beekeeper from the South Carolina coastal town of Conway called me by phone to tell me of an experience he had recently while cutting up a fallen tree. He was so convinced that this was an amazing breakthrough in apicultural research that he mailed me a cassette tape voice recording that covered every minor detail of the event.

It so happened that the strong winds of Hurricane Hugo blew down a large oak tree in his backyard a year earlier. He particularly dreaded cutting up the fallen tree because it contained a feral or wild honey bee colony in a hollow area of the tree, which is typically called a "bee tree" by the beekeeping community. The bee colony had apparently survived the fall because he could see the bees actively coming and going from the colony entrance. Being a hobby beekeeper, he had experienced some angry bee colonies over the years and observed the downed tree for months with great anxiety and fear of what would happen when he cranked up his noisy chain saw to cut up and process the tree.

The day came when he finally got up enough courage to do the job of cutting up the fallen tree that had the bee colony inside. He claimed to have selected a warm, sunny day with no clouds in sight, just a perfect day to work around honey bees. He cranked his chain saw that rang out the loud noise that he feared would set off the colony in a frenzy that would likely send him to shelter. To his utter amazement, the colony remained calm throughout the ordeal as he cut up the tree. He was extremely pleased to report that he did not receive a single sting during the entire process. He claimed the loud noise of the chain saw seemed to have a calming effect on the bees and that he was sure I would benefit by knowing this information for passing along to other beekeepers for training purposes. He passed onto me the finer details of the process including the chain saw manufacturer and the model number.

Fig. 71. Job done with no honey bee stings. (Photo courtesy of Jimmy Tinsley).

What this beekeeper did not understand is that honey bees are basically deaf and do not have the same sense of hearing as humans. The sound that honey bees "hear" is a result of movement of air particles that have to come from a short distance away within a quarter of an inch (6mm) or so. However, honey bees do detect substrate-borne vibrations in their environment and will normally quickly respond to protect their colony. Apparently, this beekeeper was able to cut up the tree without causing much vibration within the combs of the colony. Also, there can be genetic variation in honey bee colonies that controls their defensive nature when disturbed. Maybe the old beekeeper was just plain lucky that day.

Lesson learned. In this case, the beekeeper had little understanding of what activities around a colony were important in maintaining a calm environment. Honey bees do not have ears to hear like we do, so they may be largely insensitive to airborne sounds such as the loud noise created by a chainsaw. Lesson to learn here is that loud noises do not normally result in causing honey bees to get out of control or become defensive. However, if the beekeeper's activities also result in vibration of the colony by hitting a beehive with a lawn mower or a cutting string of a weed-eater hitting the beehive or causing grass clipping to enter the hive entrance, the perpetrator will likely be in for a nasty experience.

Bears and Honey Bees Don't Mix Very Well

Much of my honey bee research was conducted in the nearby Clemson University Experimental Forest which is located only a few miles from campus. The experimental forest covers about 17,500 acres of beautiful forest land with a good gravel road system with many gated entrances which makes them secure and protected from human interference.

However, one year I had an unwelcome intruder, a black bear, that entered several of my experimental beeyards and totally destroyed about 20 honey bee colonies. The colonies were part of a region-wide research project with other bee scientists from the University of Florida and the University of Georgia.

Fig. 72. Bears and honey bees collide in my research apiary. The bears won this battle.

After discovering the damaged bee colonies, I contacted the nearby bear specialist who worked for the South Carolina Department of Natural Resources. He and I rode out to the damage sites to confirm that a bear was responsible for all the destruction. From the size of the paw tracks left behind, he concluded that it was a young male bear maybe 2 years of age which had been weaned from its mother that year and had wandered down from the mountains where other larger males dominated.

My question to the bear specialist was, "Could you trap and move this bear far away from my research beeyards?" His answer was "No, because he had heard of other bear sightings within adjacent counties and he simply had no good place to release the unwanted bear." Bears are a protected animal species in South Carolina, so my other options were very limited. This is unfortunate for South Carolina beekeepers who manage their colonies in areas where bears roam freely.

This bear incident lead to me not being able to submit data for the first year of a two-year regional honey bee project. The next year I went to a lot of extra time and expense of placing battery powered electric fences around my research beeyards. As far as I could tell, the bear never returned the next year.

Fig. 73. Bumble bees and yellowjackets also feasted on the spoils of the bear-damaged, unprotected honey bee colonies. The larger bees are bumble bees and the smaller yellow striped bees are yellowjackets.

Ironically, I had earlier in my career co-authored a book chapter[13] on predator protection for honey bee colonies which covered in detail "bear protection." I now had some personal experience with bear damage and control.

Fig. 74. Electric fence placed around honey bee colonies to control bears the next year.

Another bear protection alternative is the use of a predator platform for placement of colonies beyond the reach of bears. An overhanging floor is a must on all sides to prevent bears from climbing onto the base of the platform. The major disadvantage of a predator platform other than the expense is onloading and offloading equipment from the platform floor by ladder which can be physically challenging.

Fig. 75. Predator platform for bear protection located near Cedar Mountain, North Carolina.

Bees and Swimmers Don't Mix Well Either

The one call that I always dreaded was a call from a homeowner who had a swimming pool in their yard which honey bees were using as a water source. These calls normally came in during the hot summer months when pool activities coincided at a time when honey bees need of lots of water to cool their hive. I often found myself in the middle of two feuding parties who had already reached the boiling point before I received the call.

In most cases, the pool-owner claimed the bees were coming from a nearby neighbor's honey bee colonies. The pool owner had been in contact with the beekeeper who was rarely sympathetic to their problem.

Water carrying bees simply light on a pool deck and fill their honey stomach with water which had been splashed on the deck by swimmers. Unfortunately, children in all their activities around the pool will invariably step on a bee and receive a sting which can take the fun out of a pool party, leaving the parents very upset. Fortunately, I never heard of one of these pool stinging incidents leading to a major medical emergency, although it is very possible because about 1% of our population may suffer an anaphylactic reaction to honey bee venom.

Once the honey bees are hooked on the swimming pool for their water source, it is not possible to change their flight pattern to another water source unless the pool is completely covered with a tarp or other material for several days. However, covering the pool and making it unavailable to swimmers is not a good option when the homeowner has paid a lot of money for use of their pool.

The logical solution to the problem from the pool owner's perspective is for the beekeeper to simply move the honey bee colony to another location far away from the swimming pool. This option is taken many times reluctantly by the beekeeper to keep peace in the neighborhood.

But there are circumstances in many cases which lead to the beekeeper claiming to be the victim in the outcome for various reasons. Some beekeepers claim to have lived in the neighborhood much longer than the pool owner, so why should they give up their right to keep bees on their own property? Another question that I often heard from the beekeeper's perspective was how does the pool owner know the bees are coming to his pool from his colonies and not from feral or other managed honey bee colonies in the area?

I once followed a case of this nature in South Carolina where a bitter disagreement about bees landed before the county council. If I remember correctly, other local beekeepers became involved in the case and strongly urged the beekeeper to move his bee colonies which he did, reluctantly. The point here is that the other beekeepers were concerned that the county council might approve sanctions against or restrictions on beekeeping, if the public supported such a measure. Fortunately, I did not get involved in this volatile case.

Lesson learned. It is best if a beekeeper keeps good relations with his neighbors which does require that he manage his bees in a responsible manner, so that the bees do not become a nuisance or problem to other people. The swimming pool issue is no doubt the most common problem in neighborhoods. New beekeepers should take this into account if one of their neighbors has a swimming pool. If this is the case, perhaps the new beekeeper should consider placing their colonies on another property a mile or two away from the swimming pool. The other recommendation to the beekeeper is to provide a constant water source for their bees near the beehive, making sure the bees have a landing place to take up the water to prevent drowning. The landing place could be in the form of rocks, moss or pine straw. If the bees get hooked on the nearby constant water source, they will have no need to visit swimming pools for their water needs.

Honey bees have a need for water almost year-round in some regions of the world. Water is not stored in comb for later use, but it is collected as needed. Water is used to dilute food stores by young bees when making brood food, especially when little nectar is available, such as early spring. During hot weather, honey bees use water to cool their beehive to prevent overheating. Five gallons of water are required per year for the average colony of honey bees.

Moving Honey Bee Colonies? Be Careful!

Much care should be taken when moving honey bee colonies because this is one beekeeping activity that can get out of control in a hurry. If something can go wrong, it probably will when transporting honey bees, so planning the move carefully is highly recommended.

This is one story that I will never forget. Early in my beekeeping career, I taught a beginner level short course which was hosted by the Oconee County Beekeepers Association in Walhalla, South Carolina. Most of the class members were new beekeepers who had never managed honey bees. I announced during one of the early training sessions that I had heard of a beekeeper, Mr. Ralph Tiller, located in the midlands of the state, who was downsizing and would like to sell about 85 of his colonies at a fair price. I noted that I'd give them the contact information for Mr. Tiller if anyone was interested. This was good information to these new beekeepers because honey bee colonies can be in short supply especially in spring of the year when demand far exceeds supply.

I came back to the next training session and discovered that many of the new beekeepers were anxious to get their hands on some of the honey bee colonies that were for sale. We discussed the possibility of their desires and found that about 20 beekeepers wanted 2-3 colonies each for a total of about 65 colonies. The major problem was the beekeeper selling the bee colonies lived 200 miles away and he was not planning to deliver them. One new beekeeper spoke up and said that he had a large horse trailer. Another beekeeper said that he had a dually-pickup truck with extended cab that would easily seat four people. The problem was that none of the beekeepers had any experience at moving bees and I had never moved that many colonies either. Since I was one of the few beekeepers in the class with any experience, they asked if I'd go along to help with moving the 65 bee colonies. I felt somewhat responsible for the class of mostly new beekeepers and their safety, so I reluctantly agreed to go on the trip.

After agreeing to leave Walhalla on a Friday afternoon in April, the day came for the move. The plan was to arrive at our destination an hour or so before dark making sure that all the foraging bees would be in their beehives prior to closing them up in preparation for the move. I had talked to Mr. Tiller earlier to confirm our arrival time on that day and to make sure that he would have all 65 colonies at one location. I mentioned to him that the bee colonies should have the front entrances closed up and secure with all bees inside, so that we could simply load them and return home. It sounded like a good plan to me.

We had an uneventful trip to Kingstree and met Mr. Tiller at one of his apiaries at the appointed time. When we arrived, I was surprised to see that he only had about 20 colonies at that location. Mr. Tiller claimed that he had not had time to move all the other colonies to that location and that we would have to go to two other apiaries to pickup the remaining bee colonies. Well, the delay caused by having to go to two other locations to load up delayed us about an hour in departing Kingstree which would mean we would return to Walhalla later than we had anticipated where the 20 excited new beekeepers waited for our arrival to pick up their bee colonies.

No major problem yet, however our luck ran out about 30 minutes outside of Kingstree as we were preparing to get on Interstate 20. All of a sudden, our truck lost power and gradually slowed till we pulled into a small truck stop at about 9 PM. We looked underneath the truck and a stream of liquid had followed us to our stopping point. Apparently, we had run over some kind of object that had knocked a whole in our transmission pan and all the transmission fluid had been lost which caused damage to the transmission.

On a Friday night, there was no garage in sight open in the rural area, so we were caught between a rock and a hard place. Here we had 65 honey bee colonies stored in a horse trailer with a couple of small windows for ventilation which kept air circulating in the trailer as long as we were moving. But once we stopped moving, I figured we had a problem because that many bee colonies create a lot of heat inside an enclosed area that has minimum ventilation.

Our pickup truck driver managed to call a friend in Walhalla who had a similar truck, but it would be about 5 AM the next morning before he could get to our location. At this point, we had no other choice. We opened the back door of the trailer and noticed that many of the bees had escaped their beehives and were frantically trying to exit the trailer, so we had to close the door quickly. The only thing that we could do was hope that most of the bees survived the trip. Our backup truck arrived about 5:30 the next morning.

We arrived in Walhalla about 8:30 and the excited new beekeepers were ready to take their bee colonies home. When we opened the back door to the trailer, there was a horrible mess with a mass of dead bees on the floor of the trailer and on top of the beehives. I had a very uneasy feeling about the situation because I knew the colonies had been under stress from the heat and the loss of so many bees. As I understand, about half the colonies did not survive the summer, so some of the beekeepers paid good money for some equipment. I heard that other local beekeepers came to their rescue and replaced many of the colonies that had died as a result of the ill-fated trip

Fig. 76. Correct way to transport a large quantity of honey bee colonies making sure beehive entrances are secure or the entire load is netted to prevent bees from flying (Photo Courtesy of South Carolina Commercial Beekeeper Eric Mills)

Lessons learned. Do not haul honey bee colonies during warm weather for long distances in horse trailers or other enclosed containers without proper ventilation. It is best to move them exposed, like on an open trailer or truck bed, making sure they are secure and fastened down so that hive parts stay together during the move. Securing the colonies' entrances or placing a net over the entire load is highly recommended. Moving honey bee colonies requires careful planning and coordination to be successful.

Why Paint Beehives Various Colors?

Down through the years, beekeepers have painted their beehives white for consistency and for the simple reason that white colored surfaces do not readily absorb heat from the sun compared to dark colored surfaces. This is important especially during the hot summer months. This results in the beehives not overheating as much as they would if they were painted a dark color.

However, honey bees do have a way of cooling down the beehive by placing tiny droplets of water throughout the inside of the hive and circulating air from the outside over the water to prevent overheating. This is a pretty cool way for bees to air condition their hive. Although all this super cooling is not energy efficient and cost the colony, requiring bees to collect more water and to use energy to fan their wings to circulate fresh air through the beehive during hot weather.

In the past several years, I have noticed that some beekeepers have begun painting their beehives colors other than the conventional white. Why are they doing this and does this create a problem? Are they doing this to beautify the beeyard by adding color for the viewer's sake? In most cases, the answer to both questions is no.

There is a logical reason why some beekeepers paint their beehives different colors. Honey bees like most insects can differentiate between higher colors of the spectrum such as blue, green, and violet, but they cannot differentiate the colors in the lower spectrum such as red. Beekeepers use this to their advantage when painting beehives various colors to assist the bees in returning to their own beehive.

Fig. 77. Beehives painted various colors to assist bees in returning to their respective colony.

When a honey bee flies from its parent colony, it takes an orientation flight to remember various landmarks in relation to their own beehive. This certainly helps the bee when returning home to enter the correct beehive because if she is loaded down with nectar or pollen, she will be readily accepted by any colony in the apiary. If the returning bee has nothing to offer, she will be rejected by guard bees because each honey bee colony will have its own characteristic odor. Apparently, a full honey stomach or pollen load trumps the odor factor when a honey bee attempts to enter a bee colony. So, the thinking is that if the beekeeper paints their beehives different colors this will be one additional marker for the honey bee to remember when returning to its parent colony.

Lessons learned. Now, is all this necessary to ensure that honey bees return to their own colony? Some beekeepers could care less which colony a bee enters when it returns from a foraging trip as long as it doesn't get rejected by guard bees. However, there is one practical reason why bees should return to their parent colony. We know that bee disease and pests can be spread to other colonies if much drifting takes place within an apiary. By the way, drones or male bees often drift from their own colony without being rejected by guard bees, so they too can be responsible for spreading disease and pests.

Skunks and Honey Bees Don't Mix Well Either

Modern day beekeepers should be on the lookout for skunks which can be a problem around their honey bee colonies. Skunks will visit an apiary early in the evening or night and scratch on the colony entrance at a time when guard bees let their guard down. Bees will come out to check out the scratching vibrations at their entrance and are captured and eaten by the intruding skunk. This is no surprise given that skunks main diet is insects. Honey bees are easy pickings since so many are present in one location. The same skunk will return for several days in a row for a meal and some have been seen training their young to feast on honey bees.

What is surprising is that skunks do get stung by the bees as scientists have captured bee-feeding skunks and found stings in their mouth and throat. However, the reward of a juicy meal outweighs the pain of getting stung a few times until the pain becomes unbearable.

Beekeepers should be on the lookout for signs of skunk feeding on their colonies including the presence of scratches on the hive entrance and freshly disturbed ground in front of the beehive where the skunk rolls while feeding. Another sign is to find small wads of dead bee parts on the ground near the hive entrance that the skunk has extricated.

Skunks have become more prolific in many areas now that they are not hunted for their prized fur. So, beekeepers need to be on the lookout for this bee predator because of their frequent feedings that can weaken a honey bee colony, particularly ones which are already under stress.

The best cure for skunk activities in a beeyard is to elevate the beehive so the skunk will not be able to reach the hive entrance. This can be accomplished by placement of a beehive on concrete blocks providing a clearance of about 15 inches from the ground. I have also used benches made from resting two 4x4 inch poles lengthwise on concrete blocks to get the higher hive entrance elevation for placement of beehives. Also, as a deterrent, I have seen photos of a piece of sharp metal spikes placed on the beehive entrance that irritate the skunk's feet as it attempts to scratch on the front of the beehive. In some states it may be legal to trap skunks in beeyards, however handling a trapped skunk can lead to some pretty odorous situations that may stink up the beehives and the beekeeper too.

Fig. 78. Several beehives placed on a bench to deter skunk activity. North Carolina beekeeper Edd Buchanan on left and North Carolina bee inspector Jack Hanel on right.

Best Honey Producing Bee Yard

One question that I received often from novice beekeepers was: "What kind of plants should I plant on a small plot of land for my honey bees to make lots of honey?" This was indeed a sensible question because honey production is the primary goal for most beekeepers. However, if the beekeeper only had a couple of acres of fallow land, my answer was probably a surprise to most beekeepers. Rather than spending lots of money and time planting nectar bearing plants on their property expecting big rewards, the beekeeper would be better off spending their resources on moving their colonies two to three times per year to various nectar sources to maximize honey production. The next question often came back as, "How do I know where to locate my bees for good honey production?"

The answer to this last question could be answered as a story that was shared with me by Mr. Spann Leitner, a well- known and much respected commercial beekeeper from Winnsboro, South Carolina. Spann said that he had placed some honey bee colonies in what he thought to be a very poor location one year and he did not expect much honey production. The surrounding area was covered mostly in pine trees which was not expected to be good for honey production. However, Spann shared with me that this beeyard turned out to be one of his most productive locations year after year. This came as a very big surprise to him because he could never figure out where the bees gathered such an abundance of nectar. Remember, honey bees will forage for up to two miles from their hive which offers them a potential pasturage of approximately 11 square miles or 7,040 acres (2,849 hectares). So, the trial and error method of finding good apiary locations is highly recommended, if your primary goal is to harvest lots of honey.

But, what is the secret to locating good honey production areas without having to spend time and resources using the trial and error method? You should ask other experienced beekeepers where they place their colonies, but remember they may not divulge all they know. Just as Spann Leitner did in the story above, you may have to search for good honey production locations by trial and error and try not to overlook an area just because of your low expectations. In general, the better locations will be within flight distance to rivers or swamps where significant amounts of nectar bearing plants and trees are normally found.

Lessons learned. If your goal as a beekeeper is to maximize honey production, you might consider moving your bee colonies a couple of times a year taking advantage of various nectar flows. On the other hand, some beekeeper will try to plant small acreages of nectar bearing plants, expecting to harvest a bumper crop of honey every year. Planting of this nature on a small plot of land will certainly benefit your bees, but do not expect to harvest large amounts of honey.

I experienced a good example of this on the Clemson University Forest where I had a beautiful research apiary located in the middle of a 3-acre mature yellow popular seed orchard, which I thought was perfect conditions for honey production. The apiary proved to be an excellent honey production beeyard year after year. However, most years I witnessed the bees exiting the seed orchard on their foraging trips, although the trees in the seed orchard were in full bloom. In other words, the bees preferred foraging elsewhere and ignored what I thought were perfect conditions for honey production right in their own backyard.

Honey Production in the Congaree Swamp

One suggestion to making a bumper crop of honey is to place honey bee colonies within a mile or so of a swamp. You can expect an abundance of nectar and pollen bearing plants and trees near and in a swamp where there is plenty of water most of the year to support plant life. This can also be true of an area within close proximity to a river.

A good example of this type area is the Congaree Swamp which is located in the midlands of South Carolina, just east of Columbia. I have worked alongside bee-keepers in this area who were rewarded quite handsomely most years. The major nectar and pollen producers in the area were tupelo, holly, willow, and tulip poplar.

A word of warning in areas of this nature is to be informed of the rise and fall of the water level in a swamp or river which can change rapidly. One way to approach this issue is to make sure that your colonies are accessible year-round, so that you can move them to higher ground if necessary

Fig. 79. Tupelo trees in the Congaree Swamp.

Fig. 80. Honey bee colonies in the Congaree Swamp area placed on a platform to protect them from flooding.

Fig. 81. Honey bee colonies located on hive stands on the edge of the Congaree Swamp.

on very short notice. The other option is to build a platform for placement of honey bee colonies well off the ground.

Feeding Honey Bees Too Much

I once visited an experienced beekeeper friend of mine, the late Ralph Tiller in King-stree, South Carolina, in the fall of the year and I noticed that he had a large one-half gallon feeder jar on the front porch of each of his beehives. It was not surprising to see a beekeeper feeding sugar water or corn syrup to his honey bee colonies in preparation for winter to help them get through the cold weather when no nectar is available for the bees, but the large sized jars seemed unusual.

So, I asked Ralph why he was feeding his bees with such large feeders. He explained that a nearby industry had a byproduct made of sugar that was good for feeding bees and that he had a friend at the plant who said that he could have all the byproduct that he could use free. This seemed at the time to be a beekeeper's dream come true to have an unlimited supply of free sugar syrup.

Ralph mentioned that he was going to try something new this winter since he had all this free sugar byproduct for his bees. He planned to feed his 100 or so bee colonies right through fall and winter right up until the normal nectar flows begin in his area which was probably about mid-March. I immediately warned Ralph that there was a very good chance that he was going to have a lot of swarms cast from his colonies the next spring, if he poured the feed to them as he mentioned.

Ralph continued with his plan and sure enough he called me the next spring and said that I was right, because almost every colony swarmed and that he was very tired of chasing and climbing ladders to retrieve his swarms. Now, this is not a good thing to have a colony swarm because the old queen leaves with about 60% of the worker bees and establishes a new colony at another location. The parent colony left behind raises new queens, but it takes a while for the colony to work all this out and have a single queen to start the brood rearing again.

In effect, swarming normally results in the parent colony becoming very unproductive for the year, as it has to recover mainly from the loss of most of its work force. On the other hand, catching a large swarm of honey bees just prior to or during a major nectar flow season can be a real bonus for a beekeeper. Honey production from a colony of this nature can be very productive. The receiving beekeeper in this situation is normally overjoyed, especially if the swarm comes from someone else's colony or from a feral colony. Catching swarms from my own colonies never proved to be as much fun.

Capturing honey bee swarms at the end of the nectar bearing season can be less rewarding because the beekeeper will likely have to feed the colony through fall and winter which can be an expensive process. Research has shown that swarms emerging from feral colonies have an extremely low chance of surviving till the next spring in the wild. There is a common old saying that goes something like this:

A swarm of bees in May is worth a bale of hay.

A swarm of bees in June is worth a silver spoon.

A swarm of bees in July is not worth a fly.

I suspect the originator of this old saying was probably from the Northern US. Our major nectar flows in the South begin and end about 3-4 weeks earlier resulting in earlier swarm activity, so I suggest a modification of this old saying for Southern US beekeepers:

A swarm of bees in April is worth a white pearl.

A swarm of bees in May is worth a bale of hay.

A swarm of bees in June is not worth a plastic spoon.

A swarm of bees in July is worth less than a fly.

My friend Ralph in this story had good intentions but he simply fed his bees more than they needed to get through the winter. A colony normally has only enough storage space for food reserves to last until the next spring nectar flows kick in. If the bees run out of their normal food storage space and food continues to come in, they will begin filling the brood laying area of the hive with the extra food reserves, which results in the colony not having enough space for the queen to lay eggs and begin the early brood cycle. I've heard this hive condition to be referred to as "honey bound." This lack of brood rearing space is the primary reason that honey bees swarm. For the beginning beekeeper and sometimes the experienced beekeeper, I'll repeat that again, "The leading cause of swarming is congestion in the brood rearing area of a colony." Swarm prevention and control should be a regular topic of discussion at many beekeeper meetings and remember not to feed your bees too much.

Lessons learned. Although in my experience working with honey bees and beekeepers, this idea of feeding bees too much is very rare. Normally, the opposite is true and many beekeepers are negligent in making sure their colonies have enough food reserves to get them through winter in good shape. I knew one beekeeper who claimed to never feed his bees. He reasoned that if the bees did not have enough "get up and go" about them to make enough honey to get them through winter, then let them starve. During my inspections of this beekeeper's colonies, I often found his colonies to be weak and marginally productive. I have found that the more successful beekeepers are the ones who have a good understanding of when and when not to feed their bee colonies. Too much feeding can result in swarming which can be detrimental and too little feeding can lead to colony starvation.

Should a Beekeeper Wear Gloves?

For the benefit of honey bees, there are a few good reasons why a beekeeper should not wear gloves. New beekeepers are urged to start working bees without gloves to learn how to handle frames of crawling bees which are normally in a quiet, non-defensive mode. In other words, honey bees will teach you what they do not like and if you do something that irritates them, they have a way of letting you know. Another reason is that gloves can harbor disease which can be spread from one colony to another by the beekeeper. I've read where one bee scientist claimed that sting alarm pheromone can accumulate on a beekeeper's gloves which is not a good thing.[23] The added expense of buying gloves is another good reason for not wearing gloves.

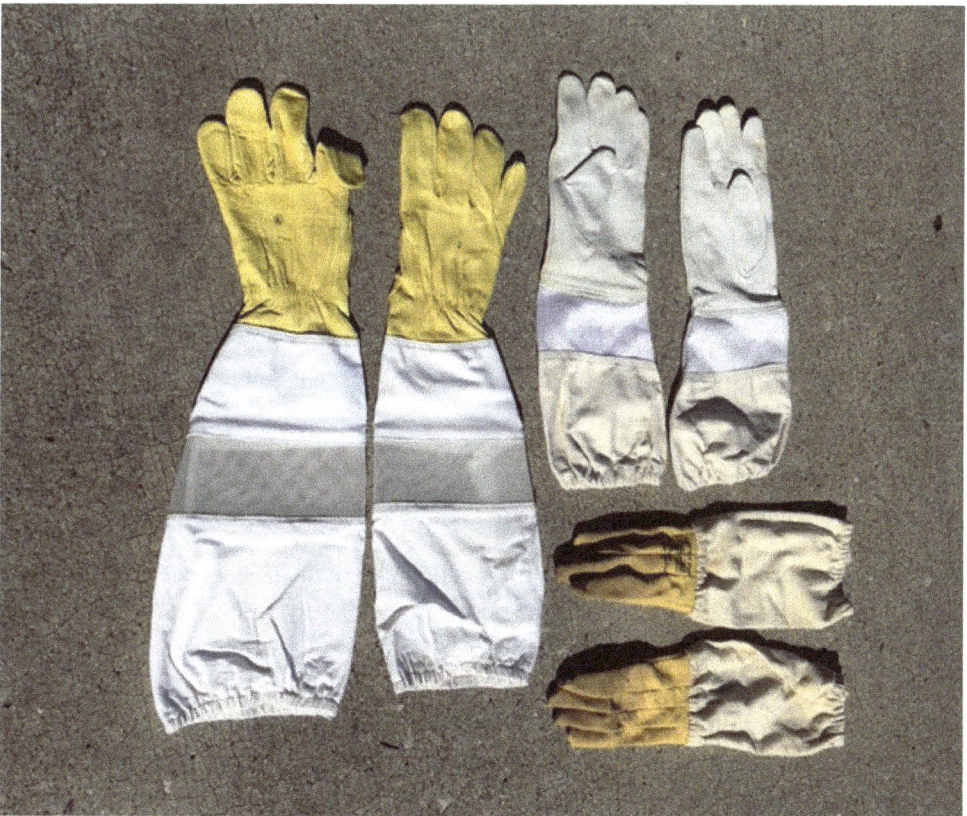

Fig. 82. Beekeeper gloves come in many sizes and are made of various materials like cowhide leather, goatskin, plastic, rubber, canvas and plastic-coated canvas.

One recommendation to new beekeepers is to learn how to properly use a smoker which tends to calm honey bees. Using enough smoke, but not too much will come with experience. Another tip is to smoke your hands prior to opening the beehive which neutralizes any foreign odors that may irritate the bees. Another good rule to follow is to not make any

Fig. 83. The author smoking his hands before opening a beehive is a good idea to mask any foreign scent that may irritate the bees.

quick, jerky moves when working with honey bees. Slow, deliberate moves are best when working around a colony of honey bees.

When a new beekeeper begins wearing gloves for extra protection, they tend to work colonies a little rougher and bang the equipment around which honey bees often dislike.

However, the beekeeper may not realize these activities irritate the bees because the gloves receive the stinger that goes unnoticed by the beekeeper. Many new beekeepers do have a tremendous fear of being stung, especially when first working with bees. But they should make a sincere effort to leave the gloves off, if at all possible, just to prove that gloves are not a necessity when handling honey bees. There are exceptions to this gloves-off rule which I have been guilty of using in the past. When I know that a colony is known for having an ill-temper, in other words prone to becoming very defensive at the slightest disturbance, I put on gloves. There are other times to wear gloves, like when a beekeeper has to work with bees in inclement weather or when the beekeeper is in a hurry to work several colonies.

Lesson learned. New beekeepers should resist the temptation of wearing gloves when first learning how to work with honey bees, if at all possible. The bees will teach the beekeeper what activities they dislike, thus making him a better beekeeper. After learning how to work colonies without getting them upset, the experienced beekeeper can put on the gloves only when necessary.

Fig. 84. A beekeeper working with honey bees without wearing gloves is recommended.

Strange Placement of Beehives

I often received questions from beekeepers and especially non-beekeepers on why I placed my honey bee colonies in such a haphazard manner, especially my Cherry Farm colonies which were on full display to the public near the university campus. Didn't I know they look much better when lined up in military fashion?

The issue here is that we humans think in terms of what we believe looks best for us, sort of like houses that are built in a subdivision with everything neat and tidy without regard to what is best for the honey bees. You will note in the photo below that I have placed the colonies around a tree rather than in straight line. Why?

Fig. 85. Honey bee colonies positioned facing different directions.

When honey bees first exit their hive and take flight, they will perform an orientation flight in somewhat a circular motion taking note of landmarks in the vicinity of the apiary in relation to their hive location. This programs their memory to return to the correct hive and not attempt to enter other beehives in the apiary by mistake. Drifting is the term given to bees that enter the wrong colony. This can lead to the spread of disease and pests to other colonies by the drifting bees. Queens returning to the hive after their nuptial flight can also enter or attempt to enter the wrong hive.

Some beekeepers claim that if they place their colonies in straight lines in military fashion that the colonies on the outer edges will have a higher population of bees over time. Foraging bees drifting to the outer-edge colonies in the apiary may result in population imbalance.

To reduce the chance of drifting, a beekeeper can place his hives a further distance apart and facing slightly different directions such as around a tree as shown above. However, facing colonies in different directions may be somewhat limited because of the recommendation to face the entrance of hives east or south. Drifting may also be reduced by planting small trees in the apiary to offer the bees other landmarks to use in returning to the correct beehive.

Lessons learned. If the goal of a beekeeper is to have foraging bees return to their respective hive, there are a few recommendations available. Placement of beehives a greater distance apart and altering the direction of entrances of adjacent hives will help. Varying the color of hives is practiced by some beekeepers and planting small trees or shrubs near colonies will offer more landmarks for the bees to remember on their return flights.

Laying Worker Colony

In a normal honey bee colony, the ovaries of worker bees do not develop and they do not lay eggs. However, one problem that a beekeeper never wants to discover is a "laying worker colony" because workers can lay only drone eggs. There is no easy solution to this problem. Many experienced beekeepers will recommend taking the bee colony several meters away from the apiary and slinging the bees from each frame in the beehive, mainly because you can't identify and eliminate the workers which are laying only drone eggs.

One of the issues here is that if the colony is only made up of mostly drones, there are few bees available to do the work of the colony such as feeding young, cleaning cells, foraging, and protecting the colony from danger. Male honey bees are simply in the colony for reproductive purposes only. With only males being produced, a bee colony is doomed.

A beekeeper must be careful when diagnosing this problem. Sometimes a queen in a colony will simply run out of sperm due to poor mating or old age and the colony fails to replace her in time. The loss of a queen following her mating flight is uncommon, but it is possible.

A few rules apply when diagnosing a laying worker colony. If a queen is not present in a colony, the bees often create a loud roar by fanning their wings when the colony is opened and the bees appear nervous on the comb. A queen will lay only one egg per cell and it will be standing on its end in the bottom of the cell for the first 24 hours. However, if workers begin laying, they will lay more than one egg per cell and the eggs will be placed randomly in the cell. The worker's body is not long enough to oviposit an egg in the bottom of a cell. If the beekeeper is diligent and catches this oncoming problem in time, in other words, before any workers start laying, he/she can simply remove the failing queen and introduce a new queen to correct the situation.

Now, let's review a brief lesson in honey bee reproduction and genetics before going further into this subject of worker bees laying eggs in a colony. A queen only mates during one brief period in her early adult life which may take only one day but can take up to 7-8 days of mating in the air with several drones which have flown into the drone congregation area. A drone congregation area is an area where drones fly and gather to mate with virgin queens. Drones can fly in from all directions from the congregation area so drones may be from several colonies within a region. This increases the odds that a queen will not mate with drones produced from her parent colony. Queens never mate inside the beehive with drones of the same colony, which reduces the chance of inbreeding, and favors outcrossing making for a healthier colony.

To produce female eggs, a mated queen simply releases sperm into the egg prior to it being laid in the bottom of a cell. The queen will have mated on her nuptial flights with 15 or more drones normally, so her sperm will have originated from drones from various colonies because the queen will mate at a distance of a mile or further from her colony. On the other hand, if a queen does not release sperm into the egg before laying, the offspring will be a drone.

Female eggs hatch and mature through the larval stage which at this point can become queens or workers depending on diet, particularly royal jelly. A female larva fed royal jelly throughout its larval stage will become a queen, whereas a female larva fed royal jelly for only the first 3 days of larval life will become a worker bee. Worker bees in a colony will normally have the same mother, but their genetic makeup can be slightly different due to the variation of sperm as derived from the queen's earlier mating with several drones.

A worker bee will not normally lay in a colony as long as there is a queen present and/or there are worker brood present. In other words, the presence of queen or worker brood pheromones in a colony suppresses activation of worker ovaries. But if neither of these are present, a worker bee's ovaries can develop and she can start laying eggs. It is possible that ovaries in 10-15 percent of the worker bees in a colony will develop and they will start laying eggs in a couple of weeks. However, remember that a worker bee never mates, therefore she has no sperm to release into the egg so all her offspring will be drones, thus having half (16) the number of normal chromosomes (32). A drone has no father, but he does have a grandfather.

Now, let's get back to this worker laying problem. The major problem here is that the other workers in the colony will often recognize laying workers as their queen and they will reject an introduced properly mated queen. A very patient beekeeper can sometimes overcome this issue by adding frames of normal brood and clinging bees from other queen right colonies. The added brood will likely produce an inhibitory effect on the laying workers allowing the colony to raise their own queen or the introduction of a mated queen by the beekeeper. This is a costly procedure taking brood and bees from other healthy colonies and possibly sacrificing a good queen. So, maybe the original solution mentioned earlier in this story and practiced by many experienced beekeepers is best after all.

Lesson learned. To avoid this problem of having a "worker laying colony," beekeepers should inspect their colonies every 4-6 weeks if possible making sure the colony is queen-right. If the beekeeper cannot find a queen, he/she should look for single eggs in the bottom of cells or look for young larvae laid consistently in comb which is evidence that a queen is present. However, if there is no evidence of a queen present, the beekeeper should place a frame or two of young brood from a queen-right colony into the queenless colony. Then, the beekeeper should introduce a new queen into the colony or allow the colony to raise their own queen if it is evident that there are plenty of drones in the area. There is an exception to this rule in early winter when a perfectly good queen may lay very few eggs for a few weeks.

If it is a time of year when drones are not present to mate with a new queen and the beekeeper cannot get a mated queen to introduce into the colony, the beekeeper's options are very limited. The beekeeper may choose to combine the queenless colony with a queen-right colony and afterwards make a split or increase when conditions are favorable. If a beekeeper is not sure what to do in this case, perhaps this is the point where a novice beekeeper should seek out an experienced beekeeper to inspect the colony and seek advice on which direction to take.

Sperm Bank Honey Bee Colony

I have found adult drone honey bees to be fascinating creatures. My first experience in observing an abundance of drones in a single colony came very early in my career one winter day. I was inspecting some honey bee colonies that had been relocated from the state of New York to our coastal Georgetown County, South Carolina for overwintering. As I opened one of the colonies to inspect it, I'll never forget seeing a colony that had large drone bees from wall to wall even some in the middle of the brood chamber. Under normal circumstances, drone do not hangout in the middle of the brood chamber but will be found mostly on the outer frames of a beehive.

At first, I was completely confused as to what kind of honey bees were in the beehive because of their large size. Then, it dawned on me that they were all drones and that the colony was doomed. I have yet to see another colony of bees with so many drones and very few worker bees present. I've often wondered how that could have happened?

One possible explanation which I highly suspect in this case was that the colony attempted to supersede in fall which was too late in the year for virgin queens to get properly mated. In other words, the drones had been kicked out of the colonies and no drones were available in the area to mate.

However, the old or failing queen was allowed to remain in the colony, perhaps running low on sperm, but she continued to lay large numbers of drone eggs only. Therefore, the colony may have been headed up by what is called a "drone laying queen." Another possible reason is that the queen was injured or killed going into winter resulting in laying workers, sometimes called "false queens." However, this theory is doubtful because of the massive numbers of drones present in the colony. Laying workers are poor, inconsistent egg layers and would not have been able to produce such a large number of drones. There must have been plenty of food stores present to support all those drones. I thought what a waste of food stores and sperm.

Another fascinating story about drone bees occurred when I had my 6-year-old daughter, Lindsey, join me one summer day at the Cherry Farm to inspect some of my honey bee colonies. I pointed out to her some of the drone bees in the colony and I mentioned that they have no stinger. She became quite proficient in finding drones, or "boy bees" as she would say, and handling them because they were defenseless. We even brought some of those drones home in a cage for "show and tell" at her school. Although many years have passed since that day, Lindsey still remembers that outing to inspect those bee colonies and play with the drones.

Fig. 86. Image shows a. drone honey bee, b. queen honey bee, and c. worker honey bee. (Photo Courtesy of Mike Bentley, University of Florida)

The average person has never seen a drone honey bee. Drones will never be found on flowers or near water because they do not forage for their needs. Drones do not even have to look for food inside the beehive as they are fed by worker bees. However, their life of leisure ends after about the 6th or 7th day following emergence from their cell because they exit the hive with a full honey stomach to fuel them on their flight to a drone congregation area which can be a mile or further from their parent colony. The drone flies within the drone congregation area with many other drones in an attempt to mate with a virgin queen. The mating process is quick and violent as his reproductive organ is ripped away and remains behind in the queen's vagina as he falls to the ground and dies. The drone's adult life is cut short if he is successful in fulfilling his reproductive duties on the first mating flight out of the colony. Only the drones which were unsuccessful in mating return home from mating flights to live another day. The only other time that drones may exit their beehive is when a colony swarms or absconds.

An additional use that we found for drones was to use them when new beekeepers were learning to handle and practice marking queens with a dot of paint on their thorax. The purpose for the dot of paint on a queen's thorax is to be able to locate her quicker when inspecting a colony. Practicing this marking technique on drones in preparation of getting a feel for marking queens was much safer and practical than using worker bees for an obvious reason.

Fig. 87. Queen honey bee marked with a pink dot of paint on her thorax making it easier to find her. You will find her in the center of this photo. (photo courtesy of Bob Bellinger, Clemson University)

An odd occurrence that sometimes show up in honey bee colonies is the presence of white eyed drones. They are a result of a mutation that appears when a recessive gene is expressed when several white eyed drones are observed in a single colony. Although white eyed drones appear to be normal inside the hive, their activities outside the colony are cut short once they emerge because they are visually impaired. Some reports claim they are blind.

Recessive genes occur in the form of mutations which show up in drones more often than workers or queens because drones are haploid receiving only half the chromosomes the female bees receive. Drones develop from unfertilized eggs and have only one parent, the queen. The mutant white eyed drones occur because recessive genes are expressed more frequently without being overridden by a corresponding dominant gene.

Regardless, the white eyed drone's navigational skills are extremely limited making them unable to fly to drone congregation areas and mate. In fact, once they exit their hive they do not successfully return which cuts their life time short. During my 30-year career of working with honey bees, I can never remember seeing a white eyed drone, so their occurrence is quite rare, but I'll keep looking.

Fig. 88. White eyed drone in center (Courtesy of Wildflower Meadows, Riverside, California)

Lesson learned. This story is not written to lessen the importance of drone honey bees. Drones do play a vital role in the success or failure of honey bee colonies. Drones must be healthy to perform the rigors of flying to drone congregation areas, finding a queen, and mating in mid-air. Without the drone's immeasurable contribution in the mating process, future honey bee colonies would be doomed.

Can a Queen Honey Bee Sting?

Although a queen bee's primary function in life is to lay eggs, sometimes up to 1,500 eggs per day, she does have a stinger and she knows how to use it. Queens have a smooth stinger, so they can sting multiple times if necessary, but normally her stinger is used only for killing other rival queens in a colony. When a honey bee colony prepares to swarm, the colony will raise about 10-12 new queens by feeding female larvae royal jelly for their entire larval life. Royal jelly is a high protein food produced by young worker bees that is fed to all larvae in the colony for their first three days following egg hatch. If a female larva is fed the royal jelly diet for the rest of its larval life which is about six days, she is destined to become a queen bee. However, if the female larva is fed only pollen and honey after the third day, she will become a worker bee.

About two days after a swarm (which includes the old queen and 60% of the other bees in a colony) exits the beehive, the 12 or so virgin queens will begin to emerge from their cells. Sometimes the first virgin queen will go through the brutal process of stinging and killing the others prior to their emergence. Queens have powerful mandibles which are used to chew an opening in the other queen cells, then she inserts her abdomen into the hole and stings her rivals to death. Other times, worker bees prevent the first virgin queen from stinging the others resulting in more than one virgin queen in the same colony. Eventually, the virgin queens will fight it out stinging each other till one survives to head up the colony.

The legendary USDA Honey Bee Research Specialist Steve Taber who grew up in Columbia, South Carolina, published a book titled "Breeding Super Bees." In his book, Steve noted that approximately 5% of colonies may have two or more queens.[23] So, how could this happen? As noted above, more than one virgin queen may be in a colony for several days after a colony has swarmed is one example. Another possible reason why two or more queens may be in the same colony is that when a colony rears more queens to replace a failing queen through a process called supersedure, there may be a period when the old mother queen is allowed to remain in the colony while at the same time a new queen is becoming established. There could be other reasons why more than one queen is found in a colony. One reason that bee-keepers do not find a second queen in their colonies is that they are so happy to have found one queen, that they stop looking for other queens. Another reason is that a beekeeper may inspect a colony only to confirm presence of a queen, so after finding a queen or evidence of her presence, there is little reason to look further for more queens.

Getting back to why queens will sting reminds me of a story told by my good friend, Mr. Laurence Cutts, who is a third-generation beekeeper and queen-breeder who has lived most of his life in Chipley, Florida. One busy afternoon in Laurence's younger days, he was collecting some new queens which had proven their egg laying capabilities following their mating flights. As he was holding the queens to load them into their individual shipping cages, Laurence realized that he had so many queens that he ran out of fingers to hold all of them, so he started holding some of them in his mouth. He must have pressed his lips together a little too hard to suit one queen and she stung him right on the lip. Laurence proved to be the only human that I have ever heard of that was stung by a queen honey bee.

<u>Lesson learned</u>. Beekeepers should be very careful when handling honey bee queens. If a beekeeper injures a queen, there is a very good chance that she will be unable to successfully perform her important duties of laying eggs, sometimes up to 1,500 eggs per day. The fact that she does have a stinger and can use it is another good reason that beekeepers should have great respect for queens.

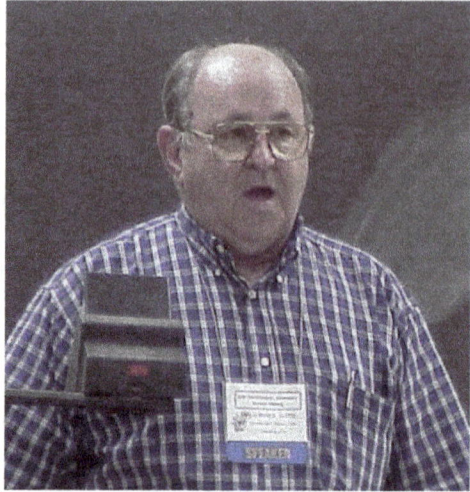

Fig. 89. Laurence Paul Cutts, 3rd generation beekeeper and queen-breeder, speaking at one of our state meetings held at Clemson University.

Good Plants for Bees

One leading question that I received often from beekeepers and non-beekeepers was "What can I plant on my property that will be good for bees and other pollinators?" This question was often difficult to answer because it will depend on the location and this may vary from state to state and region to region. My answer to this question was to contact your local agricultural agent or an experienced local beekeeper and ask them this same question.

I have mentioned in an earlier story that beekeepers should not expect a bumper crop of honey from just an acre or two of any plant. However, due to the loss of much of our nectar and pollen bearing plants in most natural settings due mostly to development, it is always a good idea to plant various trees or small plants to benefit pollinators in your area. *The goal for beekeepers and non-beekeepers is to plant a variety of nectar and pollen bearing plants that bloom from early spring through late fall.*

I'll cover some of my recommendations that have been good nectar and pollen producing plants and trees for bees in South Carolina. Some good trees for bees are tulip poplar, black gum, tupelo, willow, crab apple, maple, persimmon, and black locust. Sourwood is another good tree for bees and makes for a delicious honey, but it only produces large amounts of nectar at higher elevations. As for shrubs and smaller crop forage plants, I'd recommend camellia, sweet breath of spring, white sweet clovers, sunflowers, buckwheat, vetch, vitex, and canola, along with other plants in the mustard family. Cotton and soybeans are good nectar producers in some regions of our state. Some wild plants that are good for bees include blackberry, plum, aster, huckleberry, golden rod, gallberry, sparkleberry, privet, and dandelion.

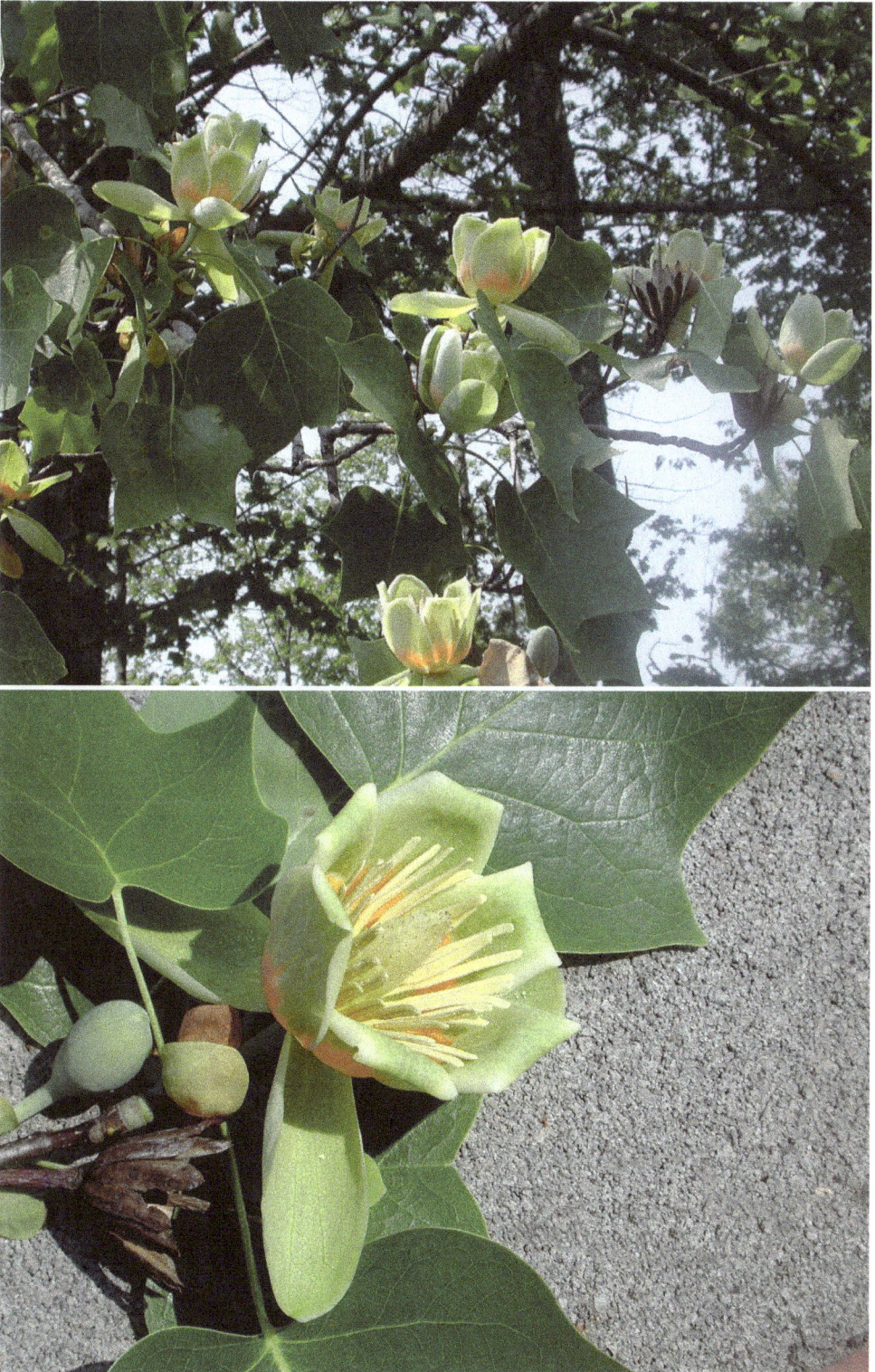

Fig. 90. Tulip Poplar tree blooms (Photos Courtesy of Paul Boone, South Port, North Carolina[7])

Fig. 91. Willow tree / Willow tree bloom (Photo Courtesy of Paul Boone, NC[7]).

Fig. 92. Tupelo trees in the Congaree Swamp / Tupelo tree blossom.

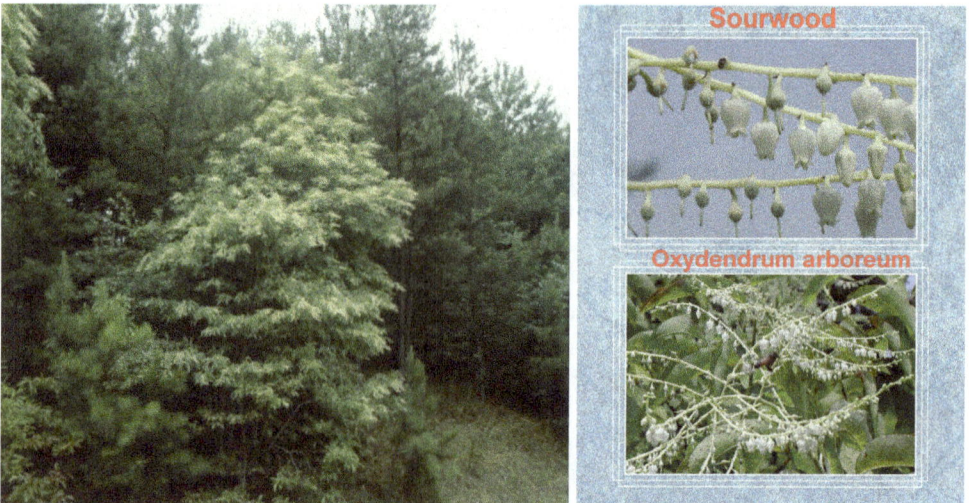

Fig. 93. Sourwood tree in Oconee County / SC Sourwood tree blooms (Photo Courtesy of Paul Boone[7])

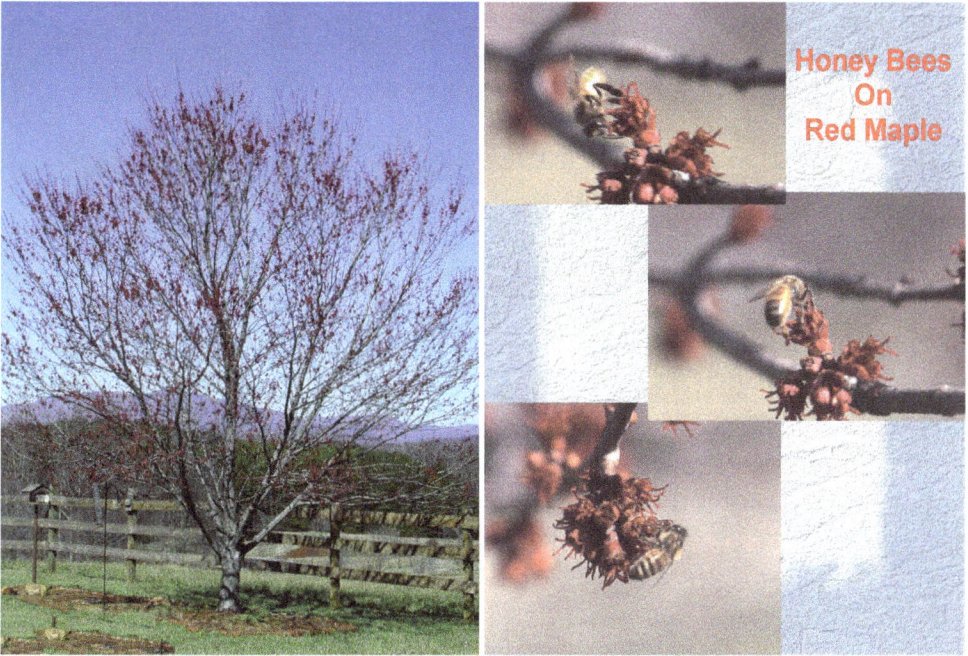

Fig. 94. Red Maple tree and blooms (Courtesy of Paul Boone, South Port, NC[7])

Fig. 95. Black Gum tree and blossom (Photo Courtesy of Paul Boone, NC[7]).

Fig. 96. Black Locust tree in bloom in May at Clemson University, South Carolina

Fig. 97. Blackberry in bloom (Photos Courtesy of Paul Boone, South Port, North Carolina[7])

Fig. 98. Beekeeper Bill Gentry in his field of sunflowers in Laurens County, South Carolina

Fig. 99. Canola in bloom

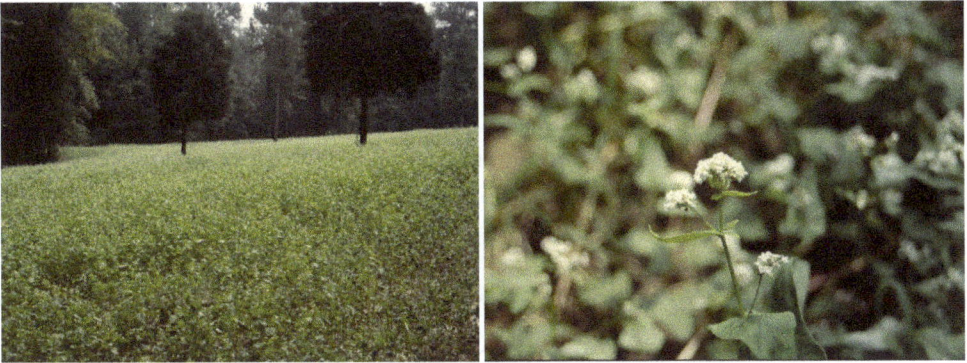

Fig. 100. Buckwheat in bloom in Saluda County, South Carolina.

Fig. 101. White Clover (Photo Courtesy of Paul Boone, South Port, North Carolina[7]).

<u>Lesson learned</u>. It is a good idea to plant various plants and trees for pollinators in your area. Pollinators benefit particularly from a variety of pollen producing plants and trees to prolong the bloom season for their yearlong nutritional needs. For better results, make sure you select varieties that grow well in your region that produce nectar, pollen or both which are beneficial to pollinators. Selecting plants and trees that are native to your area is highly recommended. Avoid planting species which may be considered invasive.

Bee Stinger Between the Eyes: Ouch!

I once had my young son, Jordan, join me one day while working with some of my honey bee colonies that were located at the Cherry Farm at Clemson University. He was about six years of age at that time, so I made sure that I suited him up with proper protection including a youth protective bee suit, veil and gloves. I knew his mother would not be happy with me if I allowed him to get stung.

As I was working with the colonies verifying queen presence and checking other details, I almost forgot about my son being present. When I looked up, I noticed that he was playing in the bed of my nearby pickup truck and he had taken off his bee veil. I returned to working with my bee colonies. Soon I heard the door of my truck close and I looked up to see my son had climbed inside the cab of my truck and he had rolled up the windows. Since it was such a warm day I found this to be quite odd, so I thought that I had better check on him. I opened the truck door and noticed that he had a bee stinger imbedded right between his eyes which I immediately removed. His eyes were watering and I knew that he had to be in pain. But to my surprise, he only had a red spot in the area of the sting site and he did not swell or have any other problems. Jordan proved that day to be a pretty tough little fellow and he never allowed this to become a defining moment. Later in life, he gladly helped me work with bee colonies.

Lesson learned. It is a good idea for everyone to be well protected when working around honey bee colonies, especially children. As a minimum, a veil must be worn because bees are naturally attracted to orifices such as eyes, ears and mouth. Getting stung on the first outing around bees as a child is not a good experience. Good supervision is highly recommended also when working with young or novice beekeepers. I was very fortunate that day that my son did not have a reaction such as severe swelling around the eyes. Apparently, the stinger must have deflected because he did not receive a severe sting. I dodged a bullet that day because his mother would not have been very pleased with my negligence in lack of supervision and protection of our young son in the beeyard.

Fig. 102. Wearing a veil is a must and light-colored clothing is recommended when working around honey bees. (Photo courtesy of Bob Bellinger).

Africanized Honey Bee Stories

Africanized Honey Bees in Upstate South Carolina

I received a call in my office one day from a young lady in West Union, South Carolina. She claimed that she discovered that she had Africanized honey bees, also known commonly by the public as "killer bees," in her backyard. My first response to her was how did she know the bees were Africanized? She claimed that she had been reading about Africanized bees and that they are often found to live in the ground, unlike our European honey bees that do not have ground nests, which is a correct statement. She informed me that the bees in her back yard appeared to be nesting under a large rock. My first thought was that these bees were yellowjackets which normally nest in the ground. After a long conversation with her, she insisted that these were honey bees and not yellowjackets.

I thought this to be an odd occurrence and that surely these were not Africanized honey bees as she strongly suspected. We did not have Africanized honey bees in South Carolina, but we had been surveying for them and highly suspected that some day they would be discovered in our state. Our strong suspicion was that they would be first discovered in the coastal region of our state near our major ports of entry and not in the upstate.

However, I figured I had better follow up and investigate. My first step was to have an experienced beekeeper, Mr. Dean Boggs, who lived in a nearby town, ride over and take a look at this suspected ground-nesting honey bee colony. Dean was a retired Clemson University Agricultural Cooperative Extension Agent and a long-time beekeeper. He definitely would know the difference between honey bees and yellowjackets.

I called Dean and he said sure that he would ride over and check out this situation. I told him that time was of the essence in this case and to be sure and call me back after his findings. Dean called me back about an hour later and informed me that the bees were definitely honey bees and that they were coming out of the ground. I told Dean to stay put at the mobile home and that I'd be there in about 30 minutes.

I drove in the yard and met the young lady as several little boys maybe 5-6 years of age were playing. She and Dean took me around to the back yard and she showed me the rock from which she claimed the bees were coming out of the ground. I did not see any honey bees as I stooped down to investigate, but I did sense a strong urine odor.

I asked her, "What was the source of the urine odor near the rock?" She explained that she had two little boys of her own and kept her sister's two little boys too. During the summer, she did not allow the boys to come inside their mobile home to use the bathroom, so the rock was their designated place to relieve themselves. Aha! Now the pieces of the puzzle were coming together. We were experiencing some very hot and dry weather the last few weeks and the boys were providing a constant water source for the bees in the form of urine as they peed on the rock. The bees were visiting the soil around and underneath the rock making it appear that they were nesting there when they were only flying in from a nearby honey bee colony, probably located in the same or nearby neighborhood. At that time, I had never heard of honey bees being attracted to urine, but it was the case here.

Lesson learned. Honey bees have a different sense of taste than humans and will often go to stagnant, brackish, or contaminated water. The lesson here is that beekeepers should provide bees a nearby source of fresh, clean water, especially in the hot summertime when bees use a lot of water to cool their hive. The water should be provided well before hot weather in order to avoid their bees seeking out other sources of possibly contaminated water or swimming pools. In the case of the above story, I have since read a couple of other instances where honey bees have been found to be attracted to urine, but I'm convinced these were isolated and unusual events. I'd say the right conditions have to occur to make this possible. Also, I am not aware of a mechanism whereby urine could have contaminated the honey produced by the bees in this story. Honey is produced by honey bees from plant nectar and it is unlikely that urine could enter the pathway of the conversion of nectar to honey.

Africanized Honey Bees Attacking Sliding Glass Doors

A lady from rural Abbeville County, South Carolina, called me one day all excited about some "African Killer Bees" which were attacking her sliding glass doors on the backside of her house. She claimed the bees were making a loud buzzing noise and were hitting her glass doors with great force. During our telephone conversation, I managed to learn that these were some very large brown bees that mostly hit on her sliding glass doors only in late evening. It seems this lady may have been watching too many horror bee movies that were playing at that time or maybe she got caught up in all the media coverage of the so called "killer bees."

As it turned out, these large bees were European hornets (*Vespa crabro*) which are true hornets that were likely nesting in a hollow of a nearby tree. Normally their colony entrance is a few feet off the ground, however these are very gentle hornets unless you deliberately upset their nest or step into their foraging pathway and provoke them. The hornets tend to fly during the day and into the evening hours and will fly into a window, perhaps seeing their reflection in the glass during their mating attempts. These are large hornets and look to be very intimidating, but they are harmless unless disturbed. They are carnivorous and will eat large insects such as grasshoppers, wasps, moths, honey bees and other large bees. So, after much discussion with this lady about these suspected Africanized honey bees instead being European hornets, she was relieved to know that the large hornets were not dangerous.

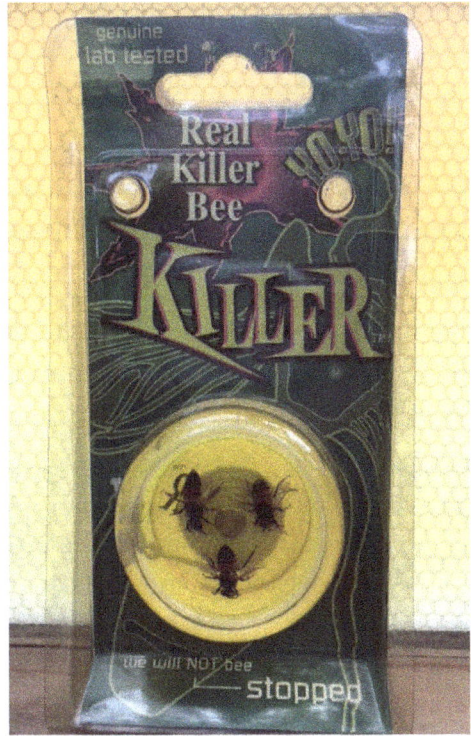

Fig. 103. Killer Bee Yoyo.

I have had some personal experience recently with European hornets which were attracted to my garage exterior light in early evening hours during the month of August. They were coming from a nest located about 50 yards from my garage and were nesting in the hollow of a cedar tree. Their flight activities were very impressive around the colony out to about 10 feet. I had to be careful not to stand in their foraging pattern as they exited and entered their colony. Unfortunately, they were nesting adjacent to my grandchildren's playground and were very active, therefore I had to eventually exterminate them as a safety precaution.

Lessons learned. European hornets are considered to be beneficial when nesting away from civilization because they feed on other harmful insects. However, the hornets should be controlled when they feed on beneficial insects like honey bees in an apiary or when their flight pattern crosses the pathway of humans or pets.

Africanized Honey Bees Suspected in Oconee County

This story occurred in the early 1990s when Africanized honey bees were continuing to make their way into regions of southern Texas and the news media was publishing stories of their progression. It seemed the public had much interests in this developing scenario and all stinging insects were suspect.

A retired corporate executive, who lived in Keowee Key, an upscale community in Oconee County, called me one day and informed me that he had Africanized honey bees in his yard. He claimed the bees attacked him and followed him into his garage stinging him repeatedly along the way. The stings were very painful and the bees were relentless and he asked me to come make an investigation into this situation before the bees spread to other properties.

I assured him that we did not have Africanized honey bees in South Carolina and that he had likely disturbed a yellowjacket nest. He assured me that these were not yellowjackets as they were dark in color and were too large to be yellowjackets. He said he had some dead bee specimens from the stinging incident and that I'd be welcome to get them identified. By this time, I figured it would be worth traveling about 30 miles to his residence and we could get to the bottom of this case.

Upon my arrival to his residence that same day, I learned from the gentleman that he had been weed-eating and happened to trim near a brush pile in his backyard and had disturbed the bees that sent him to his garage with a few stings. He reinforced the fact that individual bees followed him into his garage attempting to sting him repeatedly. The specimens that he had collected from the incident were a small species of bumble bees, which I have always heard are very defensive around their nest when disturbed and each individual bee can sting multiple times. Bumble bees are harmless and non-defensive when foraging as they work early in the day till late in the afternoon. The gentleman seemed to be relieved to learn that the bees were bumble bees and not the dreaded Africanized honey bees.

Fig. 104. Bumble Bee in Flight.

There are a few species of bumble bees that are native to the US that are found in South Carolina. Some bumble bee species are very large in size and are about the size of carpenter bees, whereas other species are about half that size. Most bumble bees have black dense-hair on their abdomen, but some species will also have yellow or light- colored hair on their abdomen.

Bumble bees are excellent pollinators for many agricultural crops and plants for wildlife, therefore they are highly valued and should be protected. Although, only a couple hundred or less individuals live in a single colony that nest in holes in the ground. Bumble bees are highly efficient pollinators as they "buzz pollinate" flowers. Their large bodies vibrate and move lots of pollen on their hairy bodies compared to other smaller bees. They do not overwinter as a colony as do honey bees and only a few fertilized queens from a colony overwinter in brush piles or other protected places. The future queens that successfully overwinter have to individually establish a new colony the following spring. A new queen's first generation of offspring will take over most of her work in the colony.

Bumble bees were present in North America long before honey bees were imported from Europe beginning in the 1600s, so they are more efficient pollinators, especially for native plants in the wild. Bumble bees do produce honey, but only enough for their own needs. Man has learned how to propagate bumble bees commercially and they are sold as expend-able pollinator colonies for various crops that are grown in greenhouses such as strawber-ries. The commercial bumble bee colonies are fully functioning pollination units upon arrival, but they may last only 8-10 weeks.

Lessons learned. Ground nesting bum-ble bees should not be disturbed for vari-ous reasons. They become very defensive when their colony is threatened and each individual bee can sting multiple times. Bumble bees are highly valued as pollina-tors and should be protected. Undisturbed-set-aside areas such as a back corner of a field are recommended whenever possible for their nesting purposes. Future queens of hibernating wild bees often overwinter in brush piles, so it is a good idea to leave some brush piles for their wintertime pro-tection. Burning all brush piles in fall may

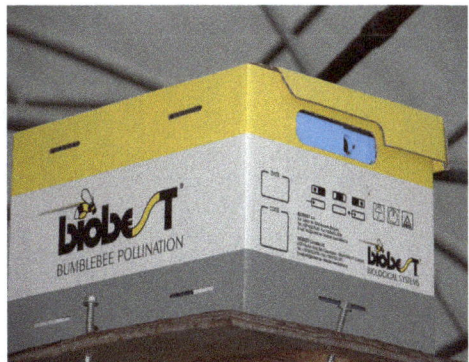

Fig. 105. Commercial bumble bee colony for pollination purposes in greenhouses.

reduce the number of wild pollinators in an area the next year. You may not win any yard-of-the-month awards by leaving an area of your back yard undisturbed or leaving a couple of brush piles during winter, but you will stand a higher chance of having wild pollinators in your area the next season.

Fig. 106. Commercial bumble bees pollinate greenhouse-grown strawberries near Easley, South Carolina.

Fig. 107. Leave brush piles like this as undisturbed shelters for overwintering bumble bee queens which will head up colonies of these valuable pollinators the next season.

Africanized Honey Bees from Arizona

I received a call in my office one afternoon from an engineer from a nearby Lockheed Martin Corp. plant which is located at the Donaldson Center, Greenville, South Carolina. The plant site was a former WWII training airbase and Lockheed Martin presently conducted maintenance on airplanes, particularly for the Department of Defense.

The engineer asked if I'd come over and check out a situation they had with a colony of honey bees. He mentioned that a large plane they were working on had arrived in Greenville from Arizona several weeks ago and had remained on the tarmac until they could get it in for maintenance.

The large aircraft had been part of the Department of Defense Star Wars Program and had been in storage for awhile on an abandoned airstrip in Arizona. The Program directors decided to activate the plane, but it needed some repairs, so the airplane was flown to Lockheed Martin in Greenville. Some mechanical problems resulted in the pilot flying the plane at a much lower altitude of 20,000 feet rather than the normal 35,000 feet.

The plane landed in Greenville in August and was stored there for a few weeks. When it came time to begin work on the aircraft, the employees noticed there were several bees flying around the horizontal stabilizer which is the fixed horizontal member of the tail assembly of the plane. The plane was taken through the airplane wash station several times to take care of the bees. However, the bees continued to fly in and out of the horizontal stabilizer.

I agreed to travel to Greenville that afternoon which was about a 45-minute drive from Clemson. When I arrived at the site, the engineer met me and I was escorted back to a large airplane hangar. As we entered the hangar containing the large aircraft, it became apparent that a pest control firm had been called earlier that day to exterminate the bees. The employees removed the outer shell of the horizontal stabilizer exposing the dead bee colony which contained fresh brood. There were several layers of comb present which indicated that the colony had been well established. I proceeded to cut and collect some of the comb from the colony and then I was airlifted to collect a few of the remaining live adult worker bees that were gathered in small groups at the highest point on the vertical portion of the tail assembly of the plane.

Fig. 108. Tail assembly of a large airplane showing the horizontal stabilizer.

After some discussion with the engineer, it became apparent that the honey bee colony had likely entered the plane as a swarm sometime earlier while the plane was in storage in Arizona. At the time of this incident, the state of Arizona was known for having Africanized honey bees which had entered South Texas in 1990 and had spread to other western states also. Since the plane had some mechanical problems, it was flown to Greenville at a much lower than normal altitude resulting in a temperature which allowed the bee colony to survive the flight in good shape. Had the plane flown at the normal 35,000 feet altitude, the bee colony would have likely frozen during the flight to Greenville.

When I returned to my office that afternoon, I shipped the adult bee sample to the Beltsville Honey Bee Laboratory in Beltsville, Maryland, which at that time was analyzing bee samples for Africanized genes. In about 4-5 days, I received word from the bee lab that the bees were indeed Africanized. I had earlier measured in my lab a series of comb cells that I had earlier collected from the suspect colony and the result fell within the African bee range. On average, Africanized bee colony cells are slightly smaller than comb cells that are built by European honey bee races. I kept this bit of information to myself until further confirmation arrived.

As far as I could tell, we dodged a bullet on this incident. During August in upstate South Carolina, we have very little nectar and pollen plants in bloom to support honey bees. If this plane had landed in Greenville in the spring of the year, perhaps in April or May, when nectar and pollen plants are in full bloom, the Africanized honey bee colony could have flourished and cast swarms over a few weeks period when the plane was stored on the tarmac at the Donaldson Center.

However, one other factor must be considered in this scenario. The Greenville area has many European bee colonies both managed as well as feral colonies which may have been able to dilute the gene pool of any cast Africanized honey bee swarms into the wild. Fortunately, luck was on our side during this incident and we will never know how this could have played out, given different circumstances.

Regardless, the Piedmont Beekeepers Association based in Greenville, in cooperation with the local Clemson Extension Agent, followed up during the next couple of years to trap for bee swarms around the Donaldson Center and to be on the lookout for defensive honey bees in the area. None were reported.

Lessons learned. The Piedmont section of South Carolina lies on the borderline of predictions of where Africanized honey bees may struggle to survive the cooler weather in winter, given the areas lack of food available to support bees for 3-4 months of the year. However, our coastal section of South Carolina is a region that Africanized honey bees could survive, given the areas mild winter weather conditions and available food resources almost year-round. Africanized honey bees do not normally store up abundant food reserves in the colony, but they expend most of their energies in production of brood which leads to more swarming than their European cousins. Beekeepers and non-beekeepers alike should be on the lookout and report overly defensive honey bees to regulatory authorities. Early detection is the key to successful eradication programs for problem insects.

Honey and Other Honey Bee Products

Propolis: A Valuable Product of the Beehive

Propolis is an interesting product of the beehive which often goes overlooked by the average beekeeper. Propolis, sometimes referred to as bee glue, is a resinous material collected by honey bees from plants and trees, particularly pine, alder, oak, poplar, elm, sweetgum, beech, blackberry, and birch. The resin is produced on buds, twigs, leaves or on areas of a tree that have been scarred to protect the tree from the elements or from fungi, bacteria, viruses and mold. Bees mix the resinous material with salivary secretions and

beeswax. Propolis ranges in color producing shades of brown, green, red, and gold. Propolis is collected by foraging bees normally in late summer and fall during the warmest part of the day. A propolis collecting bee will return to the colony and go to the part of the nest that needs propolis and wait for a house bee to unload the propolis from its corbiculae.

The antibiotic properties of propolis make it a perfect material for creating an environment for honey bees to sterilize their living quarters where thousands of bees

Fig. 109. Worker honey bee collecting propolis on oak tree.

are concentrated into a relatively small area. The bees coat all surfaces inside the beehive with a thin coat of propolis to ensure a healthy place to live. The bees also use propolis to stop up holes or crevices in the beehive and they will use it to cover or mummify any foreign object that they cannot remove from the beehive, such as a dead mouse. Sometimes honey bees use propolis to reduce the hive entrance to minimize cold air from entering in winter and to better protect against rodents from entering. I like to refer to propolis used by honey bees as being equivalent to man's use of "duct tape" because of its universal and varied uses.

Man learned to use propolis for his own various hygienic purposes beginning over 2,000 years ago to treat wounds and numerous ailments such as sores and ulcers. Propolis is a very complex material and its chemical elements can vary depending on the source. Some of the more popular honey and propolis products are produced from the Manuka tree which grows primarily in New Zealand.

Fig. 110. Propolis health products.

Several years ago, I had a York County, South Carolina, chiropractor contact me with a request for the availability of locally produced propolis because he had been purchasing propolis from a pharmaceutical company at a hefty price. He had been using propolis in some of the products that he used in his practice, such as propolis incorporated toothpaste, acne medicine and products to relieve many mouth and throat problems. I had to inform him that I did not know of anyone in the state who harvested propolis from their beehives in quantity, but I did give him contact information for some of our commercial beekeepers who might have an interest in his request. I never did hear again from this chiropractor. Since that time, I have talked to a few small-scale beekeepers who harvest a little propolis from their beehives for their own personal use.

Lessons Learned. One warning to beekeepers is that any propolis collected from beehives to be used by humans must be clean. Hive scrapings will often be contaminated with small pieces of paint or other foreign debris making them unhealthy for human use. The best way to collect clean propolis is to use a propolis trap which is made for that purpose. There are now many over-the-counter products that contain bee propolis which are available to the public.

Beeswax and Pollen

Beeswax is a honey bee product that is often overlooked too, however you will find many uses for beeswax from candle-making to use as a lubricant. Most beeswax is harvested by beekeepers as comb cappings which are removed from cells during honey processing. Beeswax has been used commercially for over 2,000 years in the cosmetic line for producing cold crème. Beeswax was used by the military up until WWII as a lubricant for pulleys and cables, mainly due to its staying quality. Other uses of beeswax by the military included waterproofing canvas tents, belts, and metal bullet casings. Some beeswax is recycled back to the beekeeping industry in the form of foundation which provides a base for the bees to build their comb.

According to the late Richard Taylor (1919-2003), renowned expert on comb honey production and author on the subject stated, "Pure beeswax, properly salvaged from honeycomb and honeycomb cappings and uncontaminated with resins, propolis or other impurities, is one of the most interesting and beautiful products of nature. Though it physically resembles other waxes in conspicuous ways, it is in fact entirely different in its chemistry, and its molecular characteristics."[24]

Fig. 111. Small candles made from beeswax

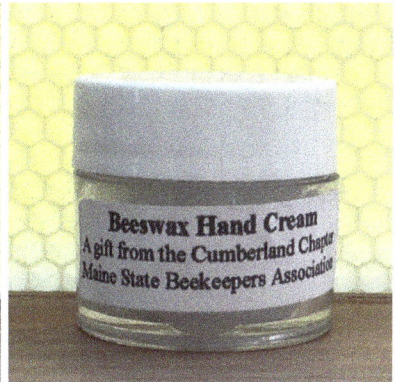

Fig. 112. Beeswax hand crème

Honey bee workers have eight wax glands underneath their abdomen that are active only in young adult bees from 1-3 weeks of age after which their wax glands begin to degenerate. The glands exude a secretion that is a liquid, but it hardens into scales on plates when exposed to air and turns into little white flakes. The wax flakes are removed from the wax plates by the bees producing them and they pass the flakes forward by their back legs to their forelegs and mandibles, where they are manipulated and added to the comb under construction. Comb building worker bees suspend themselves in what is known as "festooning" when building comb. The bees cling to each other remaining perfectly still when festooning and pass along the flakes of beeswax to a construction zone on the frame.

Fig. 113. Beeswax candles.

The worker bees manipulate the warm wax flakes into comb cells where they raise their young and store their food. A strong colony of 50,000 honey bees should be able to produce one-half pound of beeswax per day under ideal conditions.

Honey bee colonies produce much beeswax during a good nectar flow, but they will also produce wax when fed either honey, sugar syrup or corn syrup. Honey bees will consume about eight pounds of honey to secrete one pound of beeswax. So, beekeepers should be diligent in protecting their stored comb to maximize their honey production and prevent their bees from having to consume their food stores to make more comb.

I have always been amazed at how well honey bees consistently build their hexagonal (six-sided) cells. The construction goes

Fig. 114. Honey bees festooning, the art of making comb. (Photo Courtesy of Wyatt Mangum).

on exactly the same way, generation after generation without exception which is a mystery to me. The only variation is the size of the cell depending on whether it is constructed for worker cells or slightly larger drone cells.

Beeswax is made up of more than 300 components and cannot be synthesized by man. Comb made from beeswax is brittle at cold temperatures and begins to melt at 143°F. Beeswax is a very stable material and has been known to deteriorate very little after hundreds of years, if protected from temperature extremes and pests.

Pollen is a very important part of the honey bee's diet, providing their protein, fat, vitamins and trace elements. The protein content of pollen collected from various plants ranges from 12-30 percent. Worker bees forage from flower to flower collecting pollen on their hairy bodies which gathers on their corbiculae or pollen baskets that are located on their two hind legs.

Bee pollen is quite variable in its makeup, depending on the origin of the plant or tree. Collected bee pollen is not a uniform product in that a sample may be made up of dozens of different species of plants that enhances the nutritional balance of the pollen. Therefore, bee pollen collected by beekeepers may vary greatly from sample to sample.

Traps can be positioned on beehives which require the worker bees to pass through a grid that removes some of pollen from the bee's hind legs. Bee pollen can be purchased at many health food stores or directly from a beekeeper who traps pollen from their colonies. Pollen is used by humans as a food supplement or for nutritional and medicinal purposes

Raw, unprocessed honey straight from the beehive contains considerable amounts of natural pollen from many plants or trees. There are exceptions such as sourwood honey which contains very little pollen. Much store-bought honey today has been highly filtered and heat treated which results in removal of nearly all pollen and enzymes turning honey into just a calorie laden sweetener.

Fig. 115. Hi Desert 100% Bee Pollen packaged for sale by Glorybee.

Lessons learned. Beeswax and pollen are products of the beehive that can be harvested by the beekeeper. Beekeepers should be diligent in protecting comb while it is stored away from the beehive, especially from wax moth destruction. Light and good ventilation will go a long way in protecting comb in storage. Beekeepers should be conservative in trapping pollen from their colonies because pollen is a necessity for bee survival.

Red Honey? No Way!

Early in my days of working at the university, a beekeeper stopped by my office and asked me to come outside to look at something in his truck. I followed the beekeeper who claimed that he had run into something very strange in one of his honey bee colonies that was located in northern Pickens County, South Carolina.

My beekeeper friend opened the hive box in the bed of his pickup truck and lifted out a frame of capped honey that was red. He asked if I could explain why one of his bee colonies had produced this red honey? The first thing that came to my mind was that maybe there was a candy factory or bakery that used a red coloring dye in their operation and somehow the bees got into a red sugary solution and had brought it back to the hive.

The beekeeper quickly dispelled that idea because the bee colony was located in a remote section of Pickens County which was mostly forest land with only a few homes in the area and no businesses. I was at a total loss in explaining why the bees had made the beautiful frames of red honey. I told the beekeeper that I would check around to see if someone might have an explanation for this red honey.

The next week or so, I thought of calling Dr. John Ambrose who was my North Carolina extension counterpart at NC State University in Raleigh. He was one of my "go to persons" when I had a question related to beekeeping because of his vast knowledge and experiences in apiculture. Sure enough, John had the answer to our dilemma. He claimed that he had run across this a couple times in North Carolina where the US Forest Service used a red paint to mark boundary lines and to mark trees in the forest to be harvested. The forester simply paints a red band around the tree at breast high and another band at the base of the tree. After harvest, the red paint mark at the base of the tree stumps serve to confirm that the correct trees were cut.

Honey bees in all their wisdom at certain times of the year will collect a resinous material from tree wound sites or around new buds or newly emerging leaves. These resins are produced by plants to waterproof the plant or protect the site from potential invaders such as bacteria, fungi, molds, yeasts, insects and other pests. Foraging honey bees collect the resinous material and use it inside the beehive to protect them from the same maladies. Beekeepers call this plant-derived material "propolis" which the

Fig. 116. Red paint marking on tree. (Photo courtesy Nelson Paint Company[16]).

honey bee foragers collect and bring back to the hive on their corbiculae or pollen baskets which are structures located on their two hind legs.

In this case, the foraging honey bees had mistaken the red paint for tree resin and returned to their hive with the red sticky material on their feet and hind legs, tracking the red paint over the comb. Following the bees many trips to collect the red paint, the wax cappings on the honey frames took on a bright reddish appearance.

Case solved. Thanks John!

Lesson learned. The beekeeper in this story was very glad to learn the answer to his dilemma of how his honey bee colony had produced frames of red honey. The honey was not colored red, but the wax comb cappings had taken on the reddish color appearance. Of course, the beekeeper did not harvest these frames of honey, but simply allowed his bees to consume the honey to reduce any chance of harvesting contaminated honey. This was an extremely rare event for the bees to collect the red paint instead of propolis for their intended purposes.

Sugar Tasting Honey

I once talked to a first-year beekeeper who noted that his new honey bee colony had produced honey that sure tasted like sugar. The beekeeper had followed the recommended rule to always feed a startup colony of honey bees sugar syrup or corn syrup to help them get established in spring, but to stop the feeding after the bees had become established and started to forage for nectar. Normally, when a good nectar flow kicks in, the bees will prefer the plant nectar and stop taking up the sugar syrup.

However, this novice beekeeper continued to feed his honey bees sugar syrup throughout the summer. In this case, the bees continued to take up the sugar syrup perhaps due to weak nectar flows in his area. Or, the beekeeper may have started the new colony too late in the spring and his bees missed the major nectar flows.

Anyway, in the fall of the year, the beekeeper decided to harvest a couple frames of his new crop of honey since it appeared the hive was heavy with honey. Unfortunately, this first-year beekeeper harvested his sugar syrup that he had been feeding the colony all summer, which turned out to be a tough lesson to learn.

Lesson learned. Beekeepers are encouraged to order their package bees well in advance, like early winter, to ensure that their bees arrive ahead of or during the major nectar flows for their region. This allows their bee colonies to build up in spring, so they will produce enough stored honey to get them through the first winter. Feeding the new colony 2-3 gallons of sugar syrup (one-part sugar and one-part water by weight or volume) is a must to make sure they have adequate food to get them by for a couple weeks or until a significant nectar flow begins.

Apitherapy: Another Honey Bee Product

Apitherapy is defined as the use of honey bees or various beehive products for human health or medical purposes. Bee venom, propolis, pollen, honey and royal jelly are the main products associated with apitherapy. Books have been written on these hive products which include testimonials that support the benefits and value of each. Most people in the medical community claim that apitherapy falls in the realm of homeopathic medicine.

The use of honey bee venom is one area of apitherapy which receives much attention. The main pain-inducing and lethal ingredient in bee venom appears to be melittin and it is likely the component that might be responsible for much of the activity produced by bee venom related to apitherapy.

Fig. 117. Bee venom therapy: honey bee leaving her stinger behind (Photo Courtesy of Renae Ausburn)

I had one nearby practicing apitherapist, nicknamed Buzz, who I invited as a guest lecturer for my undergraduate beekeeping class at the university. The story goes that his wife had MS and had sought assistance from the medical community for years, but she had continued to suffer from the disease. Unfortunately, she had come to the point of being unable to afford further doctor visits or prescribed medications. She had become bed-ridden until her husband Buzz began the venom treatments which allowed her to at least be able to get around by herself in a wheelchair.

The day came that Buzz, along with his wife in a wheelchair, arrived in my classroom well prepared for an apitherapy demonstration. Buzz brought along a glass jar with about 25 honey bees inside that he placed in our office refrigerator for a few minutes to slow the bees' activity. After the bees were subdued, he opened the jar and removed them individually with large forceps and placed three or four of the forceps with bees intact on the table in front of my class. As the bees warmed up, they became active again by movement of their legs and wings. By this time, my students had become really curious as to what was going to happen next.

Buzz rolled his wife over in her wheelchair to the front of the classroom and lifted her blouse in the back to expose her for the venom treatments. He picked up the forceps one at a time and placed the active bee strategically on his wife's back. When the honey bee sensed the warmth from her skin, it injected its stinger, leaving the detached stinger pulsating venom into her body. Buzz's wife flinched each time she was stung during the treatment episode. Her eyes began to water because of the initial pain. The stingers were removed after about 10 minutes. After the treatment, she was able to stand for a few minutes before sitting back down in her wheelchair.

So, that was the first time that I witnessed a bee venom treatment. Buzz and his wife came back another time for a similar classroom demonstration in a couple of years. She seemed to be in about the same condition as the earlier visit. However, I contacted Buzz about two years after that for a third visit and he informed me that they would not be able to come for another demonstration because his wife's condition had deteriorated.

This all seemed interesting since at the same time I had been receiving calls from beekeepers and non-beekeepers about this subject of apitherapy. I heard of one South Carolina beekeeper who was giving bee venom treatments and charging for his services. I contacted this beekeeper and informed him that since he was charging for his services, from a legal standpoint that he was practicing medicine without a license. He took my advice and stopped charging for treatments.

The foremost issue or warning of honey bee venom treatments is the fact that a small percentage of humans can have an anaphylactic reaction to the venom which can possibly be fatal. So, a trial venom treatment by a physician is highly recommended or, if administered by a nonprofessional, a trial venom treatment should be conducted near an open medical facility.

I have known senior aged beekeepers who lost their immunity to bee venom which often ended their beekeeping careers, as a result of a recommendation from their physician. However, it is possible to be de-sensitized to a particular venom by a physician who administers increasingly larger dose injections of venom over time. I understand this series of injections requires several months to administer and the injections can be painful.

Since so many people were contacting me with questions about bee venom therapy, I decided this was a good time to provide more information on apitherapy to our beekeepers in the state. Therefore, I helped organize a symposium on apitherapy for one of our South Carolina Beekeepers Association summer meetings held at Clemson University.

We invited a couple of practicing apitherapists from Maryland to come down and help with the symposium. We also invited a local physician who specialized in allergies and who was familiar with bee venom therapy. To round out the symposium and give it some balance, we invited a well-known local family physician.

We had very good attendance at this meeting. Our symposium began with the physician who specialized in allergies. He offered some of his thoughts on bee venom therapy, especially the warning about the chance of an anaphylactic reaction to venom and the chance of a person losing their immunity to bee venom over time. He also stressed the importance of a "test venom treatment" and having an Epi-Pin available in case someone does have a severe reaction to a bee sting. He noted that all bee venoms are not the same in that someone could be allergic to wasp venom and not be allergic to honey bee venom.

The family physician was next on our program and he told a story of a female patient that had come to him seeking relief from severe pain in one of her legs. He mentioned that he tried many options including all the medications and injections that were legal and had just about given up on this case, until she came in one day and told him she no longer had the leg pain. He was very surprised and curious to know why she no longer had the pain. She claimed to have disturbed a yellowjacket nest in her backyard and was stung several times including stings on the leg where she once had extreme pain. Although he was not a big believer in bee venom treatments, he was now more inclined to think there just might be something to this apitherapy.

Our next speakers were the practicing apitherapists, a man and his wife team from Maryland. The wife had MS but she was up and walking very well during our symposium and was a serious believer in the value of honey bee venom treatments. She claimed that bee venom was an important part of her life. In fact, she had become well known where she lived in Maryland and had helped many others who were suffering from various ailments. She noted that she had them sign a form releasing her from any liability before she would give them bee venom treatments. She claimed that some days there would be people lined up at her backdoor seeking venom treatments.

That afternoon we had a bee venom treatment workshop which was well attended. I mentioned to the workshop leaders that we could not allow bee venom treatment demonstrations during the workshop on campus because I liked my job at the university and if someone had a severe reaction to bee venom, that I might be looking for other employment. I then proceeded to check on additional workshops that were ongoing in other classrooms in the same building. When I returned to the apitherapy workshop, you guessed it, the leaders had live honey bees and were demonstrating how and where to give bee venom treatments. Fortunately, they made it through the workshop without a problem.

Lessons learned. Apitherapy is known as an alternative medicine that is practiced in many parts of the world. However, precautions must be taken when administering this type of treatment to individuals. As a matter of information from various sources, I will mention that the use of a signed liability release form would not hold up in a court of law if something went badly wrong as a result of a bee venom treatment by a non-licensed medical specialist. However, it is my understanding that from a liability point of view that a person could administer venom treatments legally to members of their own family. However, treating anyone outside your immediate family could result in some major legal issues. So, beware.

Apitherapy organizations are found in many parts of the world including the American Apitherapy Association and the International Federation of Apitherapists. Apitherapy.com homepage notes that apitherapy "Promotes the science and art of using beehive products to prevent and/or heal hundreds of human and animal diseases."

Kudzu Honey?

One unexpected nectar producing plant is kudzu (*Pueraria lobata*), also known as Japanese arrowroot, which has large purple blooms that produce a sweet-smelling aroma that reminds me of grape juice. I suspect this is one plant that most beekeepers overlook when seeking out plants that produce a good nectar flow some years. I tasted

Fig. 118. Kudzu plants in bloom on 27 August 2018 at Clemson University.

kudzu honey at a beekeeper's convention held in Georgia several years ago. The honey had a reddish-purple color and tasted like grape jelly. If I remember correctly, the beekeeper who had the kudzu honey was selling it by the ounce, like $3-4 per ounce! That was the only time that I have ever seen kudzu honey for sale, so it must be rare indeed.

It does not take a rocket scientist to figure out the profit potential for kudzu honey. This would be the perfect varietal honey that could be sold at specialty shops, especially in tourist destinations. However, I suspect it would be a challenge to produce kudzu honey in large quantity that was derived from only kudzu plants to give it that distinct taste. Most kudzu nectar probably gets blended with other plant-derived nectars because other plants may be in bloom at about the same time or honey from other plants may be in the same honey super when it is extracted.

However, the potential to make kudzu honey is there for the taking. You will need a lot of kudzu to make it worth your time, say 50 acres of kudzu within a mile or so of your beeyard. It has been my experience that kudzu blooms in South Carolina (late July-September) in many areas at a time when there are few, if any other nectar producing plants in bloom. I'd recommend taking some real strong honey bee colonies, stripping them down to the brood chambers and placing them close to the kudzu, just as it begins to bloom. Place two empty honey supers with drawn comb on each hive body, cross your fingers and hope that nature will be on your side. When you first see the blooms are no longer being worked by the bees, you should remove the honey supers and extract the honey. We do the same process when we take colonies to the mountains in July to make sourwood honey. Good luck!

Kudzu makes for an interesting story in the US. It was first introduced from Japan to Philadelphia, Pennsylvania in 1876 at the Centennial Exposition as an ornamental plant and forage for cows, pigs, and goats. Kudzu was further introduced into the southern states in the 1930s and 1940s to reduce erosion and increase soil nitrogen. A program was established to reward farmers $8 per acre to plant kudzu on their land, although the US government ended the program in 1953. Over a million acres of land in the US are now covered by this plant and it has been reported to be spreading in the Southern US at a rate of 150,000 acres (61,000 ha) annually. A single vine can grow up to 60 feet per year and kudzu can overtake tall trees killing them in the process. For many landowners, kudzu is now a problem weed and they seek to control it.

Grazing animals such as goats or cows will kill kudzu in about 3-4 years. Spraying with herbicides for the same amount of time will also kill kudzu. Close mowing on a regular basis should kill kudzu in about the same length of time. Another control method for small scale areas is to spread heavy mulch over the plants.

In East Asia, kudzu is used to make various food and drinks. The roots, as well as the flowers, are used in Chinese herbal medicine. In the US, kudzu is used to feed goats and other animals on land of limited resources. Other uses include fiber art and basketry which use long runners and large vines. Another novel use of kudzu that I witnessed at Clemson University during the "First Friday Parade" held in fall yearly was about 40 fraternity brothers, some shirtless, and draped in kudzu vines that they had cut from a local field for the special occasion. From a distance it appeared like a green giant monster moving slowly down the road. No doubt, kudzu sure made for a cheap source of decorations for the parade and it was a very popular addition to the parade for several years.

Fig. 119. Fraternity brothers dressed in kudzu vines for the First Friday Parade held before the first home football game at Clemson University.

Kudzu is now considered an invasive weed because it simply grows over other plants in some areas and out competes them for limited soil nutrients. The large leaves simply block the sun from reaching other competing vegetation below. In summer, kudzu has been reported to grow a foot a day on anything that it comes in contact, such as trees, fences, power poles, and sides of buildings. Kudzu is considered by some to be the most devastating invader of the South, since General William T. Sherman marched his Federal troops through Georgia and South Carolina during the Civil War, leaving behind massive destruction.

As a ray of hope for controlling kudzu, the kudzu bug, *Megacopta cribraria*, was first seen in Georgia in 2009 near the Atlanta Hartsfield-Jackson International Airport where it likely hitched a plane ride from Asia. This was an unintentional introduction of a new insect into the US. The kudzu bug which was first thought to be beneficial spread quickly to other southern states and could be found feeding on kudzu plant leaves. However, the kudzu bug was found to feed on other plants such as soybean, wisteria, and other bean plants too.

Fig. 120. Kudzu growing over other plants and trees near Anderson, South Carolina.

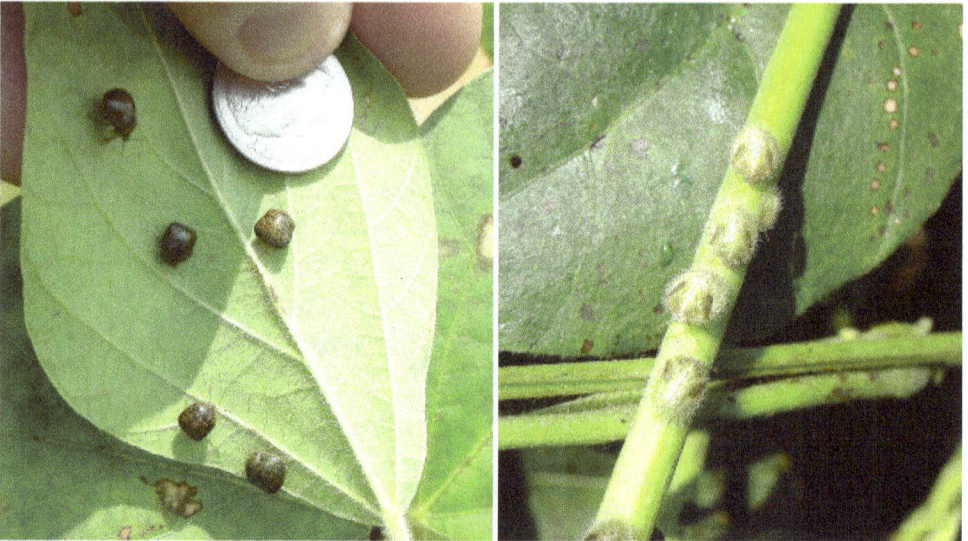

Fig. 121. Kudzu bugs on soybean leaf and stem. Adults in left photo and nymphs in right photo (Photo Courtesy of Jeremy Greene, Clemson University).

Farmers who grow soybeans estimate several million dollars of crop loss annually to the kudzu bug in the South.

Although kudzu bugs are most active in spring and summer feeding on plants, they can become a nuisance to homeowners in winter as the insects seek shelter in cracks and crevices of buildings. The kudzu bugs give off a foul odor which is attributed to their secretions, as they are closely related to stink bugs. I remember once collecting several kudzu bugs in Georgia and in the process of handling them some of the bug's secretion found its way from my hands into my eye which proved to be very painful. So, be careful if you handle kudzu bugs.

On another occasion, a gentleman drove up to our lab at the Cherry Farm in Clemson one day and asked us what kind of bugs were covering his car. Sure enough, it appeared hundreds of kudzu bugs lined the rim of his front and back windows, which proved to be a strange site. The man did not have a clue as to what was going on. Apparently, a nearby farmer had harvested soybeans in the area and the kudzu bugs having no foliage to feed on, flew to his property in mass. I have no idea why they were attracted to his car other than perhaps it being a tan color, that I understand kudzu bugs are attracted to light colors.

Fig. 122. Kudzu bugs on white car. (Photo courtesy of Lisa Dimeo).

Mysterious North Carolina Blue Honey

A mysterious blue honey occasionally shows up in the Sandhills of North Carolina. Some beekeepers claim the blue honey has been showing up within a 50-mile radius of Fayetteville, NC in beehives for over 100 years. In the jar, the blue honey looks almost like a dark purple, but when spread, it has a slightly blue tint. After the blue honey ages, it turns brown and loses much of its sweet taste.[17] For some North Carolina beekeepers, harvesting the blue honey is like striking gold because it is somewhat unusual and likely brings a premium price. The production is somewhat unpredictable and adjacent beehives may vary greatly in the presence or absence of the blue honey.

Fig. 123. North Carolina blue honey (Photo Courtesy of Lissa Gotwals).

There are a few varying explanations to the origin of such an unusual honey. One theory is that the blue honey is produced by honey bees that forage on berries of huckleberry plants whose berries are present in July and August in the North Carolina Sandhills. This theory is heavily supported by former state apiary inspector the late Bill Sheppard who was a well-respected beekeeper and a famous "go-to-guy" for answering beekeeper's questions in his home state of North Carolina. He claimed the blue honey originates when the honey bees feed on the juices through slits made on the surface of the berries which were made by other insects.[17]

Other beekeepers in the state claimed that the answer to the origin of the North Carolina blue honey production secret was unlocked in its taste. The after-taste of the blue honey leaves behind a fruity taste, similar to blueberries. So, there you have it as huckleberries are a distant relative to blueberries.

However, another theory of the origin of the blue honey was suggested by the late Dr. John Ambrose (1944-2015) who was the North Carolina State Apiculturist at NC State University for many years. Dr. Ambrose conducted a research study in the 1970s to nail-down the origin of the blue honey. He claimed the berry theory was incorrect for the production of the blue honey because berry juices are either white or pink, but are not blue.

Dr. Ambrose theorized that the blue honey was a result of excess aluminum in some of the flowers the bees were foraging. He claimed that plants in the Sandhills were taking up more aluminum from the soil than normal, compared to plants in other parts of the state. The aluminum showed up in the flower nectar which was collected by the foraging bees. The higher aluminum content nectar was reacting with the acidity being added by the house bees as they chemically converted the nectar to the blue honey. Dr. Ambrose theorized that acidity plays an important role in the production of the blue honey.[17] He collected several flowers from plants and soaked them in the bee's digestive contents overnight and found one specific flower, sourwood, was slightly blue.[3]

Dr. Ambrose had earlier tested the honey stomach contents of foraging bees returning to the blue honey hives and never found the blue nectar in the bee's honey stomach. However, the honey bees leaving the beehive had trace amounts of the blue honey in their honey stomachs. Therefore, Dr. Ambrose concluded that the conversion and production of the blue honey was taking place inside the beehives.[17]

Fig. 124. Sourwood honey blooms (Photos Courtesy of Paul Boone, South Port, North Carolina[7]).

Lesson learned. But, here is the bottom line from my perspective. Honey bees cannot forage and collect fruit juices, soda juices, or sugar water and convert them into honey. Honey can only be produced by honey bees which chemically convert plant nectar into honey. Man has never been able to produce honey without the help of bees. Only honey bees and bumble bees can successfully convert nectar to honey. So, I'm siding with Dr. Ambrose's theory until more convincing evidence supports another theory.

Foul Tasting Honey

A representative of an Atlanta, Georgia based company that produces educational materials for schools called me one day and asked if I'd like to help produce a video titled "How Honey Is Made" onsite at Clemson University? I decided to take them up on the production, so they rolled into town one morning. The crew included a project manager, a photographer/cameraman, and a free-lance actor. It proved to be a very interesting experience working with a crew that had never been around a honey bee colony. Although, by the end of the day we had produced an impressive video.

Part of the video was produced by placing a micro-camera down inside a beehive that showed bees at work inside a beehive. Another scene, which I played a role, was having the actor working with me while opening and removing a super of honey from a colony of bees. The actor made great effort in making the video entertaining for the intended children audience. We processed the honey that day in my bee lab and videoed the entire process including filling jars with the honey.

This happened to be in the fall of the year which is a time that I rarely harvest honey for human consumption because the honey gets contaminated with a foul-tasting nectar from white aster which is a wild plant that grows abundantly in our region. Novice beekeepers will often suspect a problem in their apiary during a strong aster nectar flow in fall, because the foul odor emanates from the beehives where the bees are converting nectar into honey. To my knowledge, aster honey is not toxic to man or bees and I have never heard of anyone becoming sick after eating it.

The bees don't mind the foul taste of the aster honey, so I'd leave the honey in the beehive for their winter consumption. We had just gone through a couple of weeks of a good aster nectar flow before producing the video, therefore I suspected the honey that we had just processed as being the nasty-tasting honey. At the end of the day, I gave the video production crew several jars of honey, which of course, was some of our delicious spring honey instead of the aster honey.

Fig. 125. White aster in South Carolina on left and purple aster in Tennessee on right.

I had a few jars of this nasty-tasting honey left over following the video production. Now, what was I to do with the remainder of this foul-tasting honey? I decided to run a test on one of the jars of honey. An acquaintance at the university and his wife were what I call honey connoisseurs as they were always interested in trying new varietal honeys that I provided them free of charge. They were always willing to give me good feedback on how they liked different honeys. So, you can imagine what I did with this jar of white aster honey. Of course, I presented it to them and reminded them I'd love to get their feedback.

I was sure they would come back with some negative comments about this nasty tasting honey. I was dead wrong as they reported to me that this aster honey tasted good. Boy, was I surprised. I've given this incident some thought over the years, and I've concluded that anytime you give someone honey free of charge, that most people will go out of their way to compliment it, regardless of how bad it may taste.

Another nasty tasting honey that I have encountered in upstate South Carolina includes honey produced from mountain laurel, *Kalmia latifolia*. Mountain laurel grows in the higher elevations in South Carolina and blooms at about the same time as sourwood. If we have a poor sourwood nectar flow, the bees may forage on mountain laurel instead, making a very bitter tasting honey. The mountain laurel honey is beautiful in the jar, but it will sure give your taste buds a fit and you will go looking for a place to expectorate it immediately. I've read that highly concentrated mountain laurel honey is toxic and could make someone ill, but I honestly cannot see how anyone could swallow it. I've run into some very disappointed beekeepers who rushed to harvest their supposed sourwood honey, later discovering that it was the bitter tasting mountain laurel honey. The honey was beautiful in the jar, but they regretted not having tasted the honey before they extracted it and filled their jars.

Fig. 126. Mountain laurel in bloom (Photo courtesy of Paul Boone, South Port, North Carolina.)[7]

So, what is a beekeeper to do if he encounters a bitter taste in the honey that he has harvested? I've heard from one commercial beekeeper that he blends the bitter tasting honey with other honey, say 80-90% good tasting honey, and the resulting taste is fine. The other option is to allow the bees to consume the honey which is a good alternative.

Of the countless number of other plants and trees that bees visit, comparatively few are known or suspected to produce nectar or pollen that is toxic or poisonous to man or bees and their brood. In most cases, bees will forage on many plants in any given area which will dilute the effects of the injurious plants.

There are other plants that grow in the Southeastern U.S. that produce poisonous nectar or pollen that are suspected of having a detrimental effect on honey bees. Our South Carolina state flower, yellow jessamine, *Gelsemium sempervirens*, produces toxic pollen when collected and fed to bee larvae will result in dead brood. The removal of pollen concentrated combs is highly recommended to minimize the long-term effects of the toxic pollen. Adding brood and bees from healthy colonies to the affected colonies is advised. Fortunately, other preferred nectar bearing plants and trees attract pollinators most years in the coastal region of our state where yellow jessamine grows naturally. I have never heard of a few plantings of yellow jessamine by man as having a negative effect on bees in upstate of South Carolina, where the plant does not grow naturally.

Fig. 127. Yellow Jessamine, South Carolina State Flower, growing wild in the midlands of South Carolina.

Summer titi or leatherwood, *Cyrilla racemiflora*, and rhododendrons, *Rhododendron* spp. have been reported as being poisonous to bees, also.[5] The poisoning of bees from plants is normally limited to small areas or patches, so beekeepers should avoid these areas during certain times of the year when suspect poisonous plants are in bloom. It is very rare to find concentrations of rhododendron plants in a single area in the U.S. for honey bees to make this toxic honey.

There are records of toxic honey, often referred to as "mad honey," being so potent that ancient armies used it as weapons of war, and sometimes quite effectively.[10] According to Adrienne Mayor, a research scholar in classics and history of science at Stanford University, "The ancient Greek commander Xenophon, who led his army of 10,000 soldiers from Persia to Greece in 401 BC, prided himself on choosing healthy and safe campsites in hostile territory. Xenophon noted nothing unusual about the campsite in Pontus, on the Black Sea coast in Northern Turkey, but he did note an extra ordinary number of swarming bees and said that his men soon discovered the hives and gorged on the sweet treat of wild honey. Xenophon recorded his thoughts for posterity and was appalled when his soldiers suddenly behaved like crazed madmen and collapsed *en masse*, says Mayor. His entire army was paralyzed and incapacitated for days, totally vulnerable to possible enemy attack." However, Xenophon's army recovered before they were discovered and slain.[10]

Fig. 128. Rhododendron blossom. (Photo courtesy of Paul Boone, South Port, North Carolina.)[7]

There is a later example of man using mad honey in warfare against the invading forces of "Pompey the Great" which occurred in the same region that Xenophon's army had traveled and eaten the toxic honey earlier. In 67 BC, the local Mithradate soldiers in Asia Minor retreated, but left behind pots of toxic honey which modern historians believe that the honey was produced from honey bees foraging on rhododendron (probably *Rhododendron ponticum* or *Azalia pontica*).[2] The invading Roman soldiers engorged themselves with the toxic honey which produced intense sickness and hallucinations in Pompey's army, making them incapable of combat. The local Mithradite soldiers returned and ambushed the Romans killing 1,000 soldiers with few local casualties.[4]

The use of mad honey was used again during the US Civil War when Union troops invaded a mountainous area where they discovered beehives full of honey which they pilfered and engorged on the honey. The unit became sick and disoriented immediately making them unfit for combat. The toxic honey may have been produced from honey bee colonies that had foraged on mountain laurel.[8]

Lessons learned. In most cases, if you "give" a person honey, they will come back with a positive comment, regardless of the taste. However, there could be exceptions to this rule. Aster honey is a foul-tasting honey, but it is not in the same league as the bitter tasting mountain laurel honey which is considered toxic and can make someone ill. There are other plants that can be poisonous to man or bees which the beekeeper should become familiar.

Define Local Honey

L ocal honey is honey that has been produced by honey bees in an area where the same nectar and pollen producing plant and tree varieties are found, normally in a similar geographical region. This could be a large area or it could be a very small area, given that the nearby surrounding geographical features may change, resulting in change of plant and tree varieties. All honey has pollen grains that are derived from the nectar and pollen bearing plants and trees in a given area. The only exception is that honey bees sometime collect sugary secretions from other insects such as aphids, but the secretions are still plant-derived.

Many consumers prefer to purchase local honey for the medicinal value with the popular belief that eating local honey is a natural remedy for their allergies and asthma. Some people begin taking small doses of locally produced raw honey in advance of the allergy season with the intention of building a natural tolerance to the effects of local pollen. To my knowledge, there is no scientific evidence to support this claim. However, there have been many anecdotal reports of success from people who were convinced that eating small doses of locally produced honey helped them control their allergies.

I am often asked the question, "Where can I buy locally produced raw honey?" My answer to that question is to find a reputable beekeeper who manages honey bee colonies in your immediate area, processes their own honey and sales directly to the consumer. The nearer the source of the honey the better. In addition, the honey should not be highly filtered or heated. The consumer's intent is to buy honey which has all the natural ingredients intact, including pollen.

Many honey consumers would rather pay their money for honey that contains local pollen as a bonus. Honey produced from plants and trees that are dissimilar to the ones in their area will not have the same pollen makeup. Buying honey that has been highly filtered and heated (pasteurized) is not recommended for someone that intends to use the honey for its medicinal purposes.

Normally, honey that is sold in big box grocery stores has been heated and highly filtered to remove all pollen and yeast particles to extend the shelf life and give the

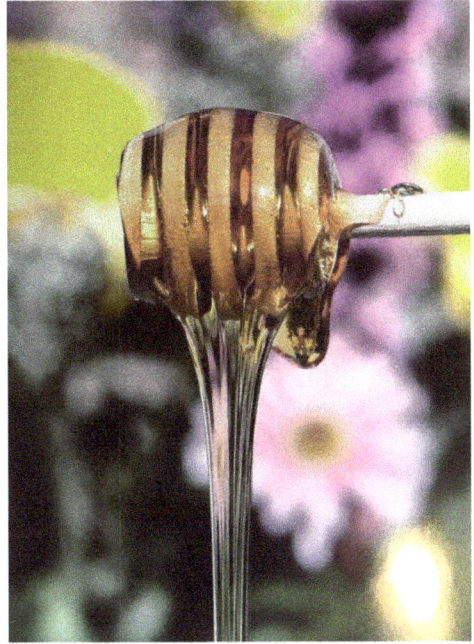

Fig. 129. Beautiful honey (Photo courtesy of John R. Steedly, Naturecraft Photography, Lexington, South Carolina).

honey a crystal-clear appearance. The average consumer prefers this beautiful honey that is still honey and it maintains its sweet taste, although much of the pollen, enzymes, and vitamins have been removed. Normally "local honey" is not sold in big box stores or large grocery stores.

Another issue that comes to mind regarding honey is: How do you know the floral source of honey? Most beekeepers will play it safe and say that their honey comes from a mixture of plants or trees in their area. Other beekeepers will claim their honey is *unifloral* or that it comes from one floral source. Officially, at least 51% of the honey must come from a single source plant or tree to be designated *unifloral*. The only sure way to identify the floral source is to test the honey for its pollen content. Each floral source nectar has pollen grains of a unique shape that can be identified only under high magnification. Few beekeepers can afford the cost of laboratory analysis to confirm the nectar source of their honey. Therefore, the term "mixed flower honey" is commonly printed on honey labels.

Several years ago, I became curious to know the floral nectar content of the honey which was produced by my university bee colonies in the spring. I was somewhat surprised to learn that the laboratory results of my honey that I harvested in June included 14 different nectar source plants or trees. The highest content floral source was about 30% from the

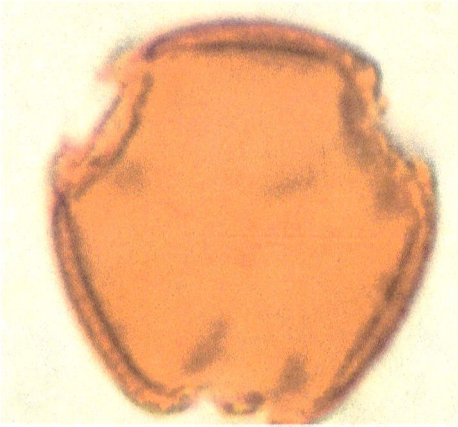

Blackberry

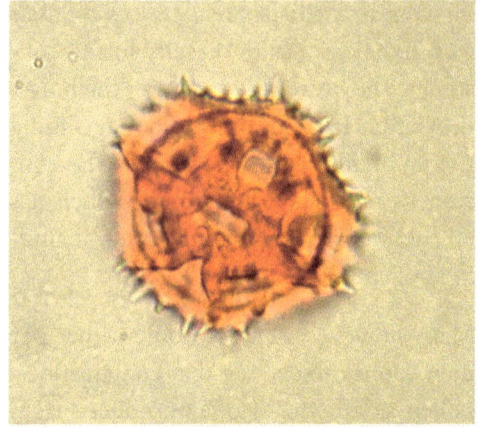

Dandelion - Blowballs

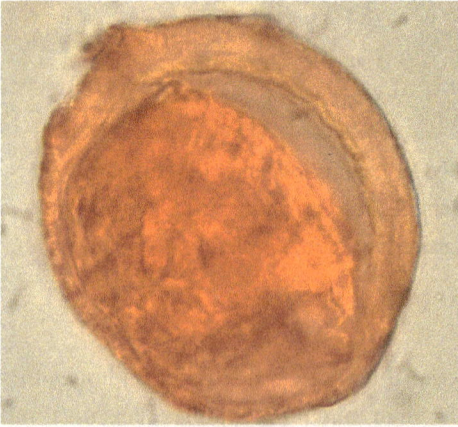

Tulip poplar

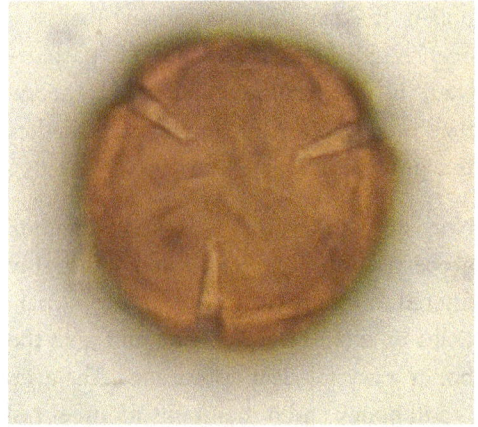

Sourwood

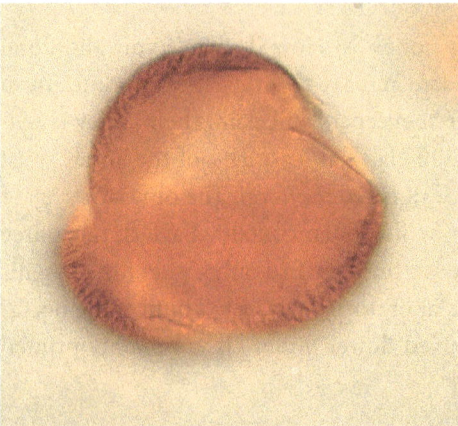

Red Maple

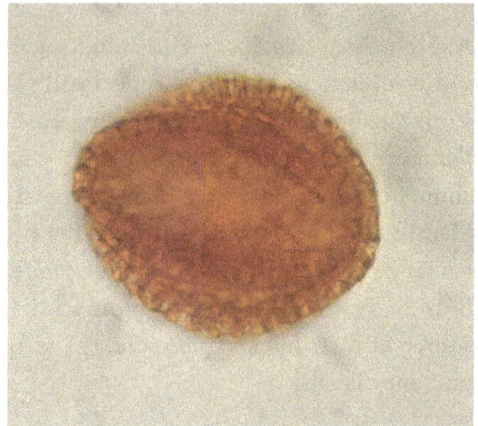

Privet Hedge

Fig. 130–133. Pollen grains of some common plants and trees (Courtesy of Paul Boone, South Port, North Carolina[7]).

Rosacea family that I assumed was blackberry which grows quite abundantly in the Clemson area. Therefore, my local honey was considered to be "mixed flower" since there was no dominant floral nectar source found in it of 51% or greater.

I tried to always clear out all my harvested honey by the end of December because that is about the time of year that it began to crystalize or turn to sugar. The more formal word for this process is to "granulate." Honey is a super-saturated sugar, mainly glucose and fructose, and cannot be held in suspension for long periods of time. Honey with higher concentrations of glucose will granulate faster. Most honey will not granulate for about the first six months following extraction, but there are exceptions, such as canola honey which will crystalize sometimes within a week following harvest. Large commercial honey processing operations flash heat honey to slow down the crystallization process.

On a small scale, granulated honey can be re-liquified by loosening the jar lid and placing the jar into a water bath having a temperature of 95-120°F for about 30 minutes. I do not like the idea of placing a jar of granulated honey in a microwave oven to re-liquify it. It is easy to overheat the honey in a microwave oven which gives it a scorched taste and the microwaves may affect some of the natural ingredients in honey.

When re-liquifying honey in plastic containers, you should be more careful and not overheat the container because plastic will tend to sag and lose its shape.

Another one of my favorite ways to re-liquify granulated honey is to place the container on a window seal or on a hard surface exposed to sunlight. The heat from the sun will do the job. This process may take several days, but if you are a patient person, it may work for you too. My attempts to re-liquify granulated comb-honey have not been successful as the wax often melted during the process resulting in a mess. Feeding the granulated comb-honey back to a bee colony as a recycling measure is a good alternative.

Honey granulation can lead to a more serious problem of fermentation or souring of the honey. Once honey ferments, there is no way to reverse the process. To prevent honey from fermenting, it is advised to not extract honey that has a moisture content greater than 18.6%. However, if a beekeeper finds himself with a batch of extracted high moisture content honey, he can remedy the situation by blending it with other honey of lower moisture content that gives a combined 18.6% or less.

If honey is going to be stored for long periods of time, it is recommended that the honey moisture content be approximately 17% or less to prevent fermentation. However, honey can be stored indefinitely, although as it ages, it will turn darker and lose its natural taste.

The moisture content of honey can be checked easily with an instrument called a refractometer. It is advised not to leave honey containers open and exposed to the atmosphere, because honey is hygroscopic which means that it will absorb moisture from the air that can result in fermentation.

Fig. 133. Refractometer, an instrument used for measuring the water content in honey.

There is one honey floral source that is extremely slow to granulate and that is sourwood which is a premium grade honey that is produced in the higher elevations of the Carolinas. I have stored a jar of pure sourwood honey in my office for eight years without any granulation. However, if sourwood honey is contaminated with other floral source nectar, such as sumac which blooms at about the same time as sourwood, it too will granulate in a few months. This is how I was able to roughly determine if I had pure sourwood honey or not, by how long the honey remained liquid.

Lessons learned. There continues to be a demand for locally produced honey, but buyer beware when shopping for it on the open market. The best way to find raw, unprocessed honey is to find a local beekeeper in your immediate area who processes his or her own honey. Or, a better way is to become a beekeeper and produce your own honey. Be careful when buying honey in big box stores which the honey has often been heated and highly filtered which removes some of the natural ingredients including pollen. Granulated honey or sugared honey is still good honey that has gone through a natural crystallization process, because it is a supersaturated sugar product. Controlled heating the honey will return the product back to its original liquid state. However, I have known people who preferred granulated honey as a spread on toasted bread. As long as the moisture content of honey is 18.6% or less, it will not ferment or go bad. Although, aged honey after a year or two will tend to lose its original taste and will turn darker in color.

Fig. 134. Sourwood comb honey, no crystallization after eight years!

The Honey Moon Is Over

During my career at Clemson University as extension and research apiculturist, I had the overall responsibility of managing sometimes up to 120 honey bee colonies with the total number varying each year. The colonies were used to train new beekeepers, to teach an undergraduate level beekeeping course, to offer field days at the university's apiaries to visiting groups and classes, and to conduct honey bee research which sometimes committed up to 75 colonies on a single project. Many of the multi-use colonies produced a good crop of honey which had to be harvested each year.

Much of the honey that was harvested was donated to the Entomology Graduate Student Club for their on-campus sales to employees. Each year the club collected several hundred dollars through their honey sales. It became a very popular event in the fall of the year for university employees to be able to purchase locally produced honey for their consumption and gifts for others. At the same time, the employees knew that the funds were going to be used for a good cause. The students used the funds to host departmental cookouts and to defray some of the graduate student travel costs to attend professional meetings.

I also took the liberty of giving away lots of honey to many friends and visiting guests of the university. Honey is a very nice gift, especially to people who enjoy natural food products and to many consumers who claim that honey helps with their allergies. A nice container of honey makes for an excellent Christmas gift, especially if the receiver knows the origin of the honey. I had a habit of making rounds throughout the university in mid-December leaving behind containers of honey to offices such as the President of the University. I know that it was well received by many because I often received a nice thank you note from the recipients. Also, I made sure that my family and close friends received a jar of Clemson honey at Christmas time and at other times of the year. The gift of honey is universal in its presentation as being a very special favor to the recipient, especially if it comes from the beekeeper. Honey also makes perfect gifts or treats for weddings or other special events.

All this past generosity of honey became an issue after I retired, because I no longer had the honey bee colonies to generate large quantities of honey. Most of my acquaintances who had received free honey from me in the past continued to have a desire for locally produced honey. When they asked me for honey, I simply told them, "The honey moon was over for this old boy," so you will have to find you a local beekeeper and likely pay for your honey from now on, or better yet become a beekeeper and produce your own honey.

Fig. 135. Small jars of honey for wedding treats.

Fig. 136. Annual harvest – beautiful honey.

Bonus Stories
(Tall Tales)

Up until this point, all stories have been true in this book. As a bonus, I have taken the liberty to include three "tall tales" which I have heard in the past. One of our standing traditions at our annual South Carolina Beekeepers Association summer meeting for several years was to hold a "Tall Tales Contest" during our Friday night dinner and social time. The first and second place winners of each contest were presented trophies for their efforts. Now, I can assure you that I heard some pretty big tales during those contests over the years, so I thought it would be a good idea to share a couple of the stories which won first place trophies.

NOTE. Remember, the following stories are tall tales and there's probably not an ounce of truth in them. However, I hope that you will enjoy these tales as much as I have over the years.

Fig. 137. Steve Forrest, Owner of Brushy Mountain Bee Farm, Telling a Big Tall Tale.

My Pet Catfish

My dad was a serious fisherman and one day he brought home a large catfish which he placed in a large tin tub in our backyard to keep it alive till he had time to process it. It just so happened that at about the same time my pet dog had died and I was a very lonely young lad. So, I asked dad if he would give me the big catfish to become my new pet?

Dad responded that he thought this was one crazy idea saying that he had never heard of a pet catfish. I said, please dad, I need a pet to replace my dog and I think the catfish would be a perfect pet. He agreed reluctantly to give me the big catfish.

Well, I soon found out that fish do not breathe too well when out of water. So, I placed him on a training program and I would take him out of the water for short periods at first, just a few seconds at first. The fish would huff and puff at first until I replaced him back in the water. I was able to increase his time out of water and he would still huff and puff, and huff and puff until I replaced him back in the water. Over time, I was able to extend his time out of the water and finally he learned to breathe without all the huffing and puffing.

One day a few of my friends came walking down the road, each one with his dog on a leash walking behind him. I asked dad if he thought I might be able to teach my pet catfish how to walk. He said that if I taught my pet catfish to breathe out of water, that I should be able to teach him how to walk. So, I set about teaching my catfish to walk. Catfish have a forked tail and my catfish was able to learn how to strut behind me on his tail. Next time my friends came walking by, I placed a leash around my catfish's neck and we joined them.

As we all walked down the road with our pets behind, we came to a pond. My friends started throwing sticks in the pond and the dogs would retrieve the stick for its owner. One of the boys asked me if my catfish could retrieve a stick out of the water? So, I thought for a moment and reasoned that of course my pet was a fish and that surely, he could swim and retrieve a stick. I threw a stick in the water and threw my pet catfish in the pond, and you might have guessed it, my pet catfish had forgotten how to swim and drowned!

The Three-Legged Chicken

In my earlier years during my career at Clemson University as the State Apiarist, better known as the state honey bee inspector, I had the responsibility of heading up the state-wide Varroa Mite Survey. Varroa mites, a major honey bee pest, had not been found in South Carolina, but we knew their arrival was imminent.

My work early in the survey was confined mostly to commercial beekeepers who were known for moving bee colonies interstate. One of those commercial beekeepers was Huck Babcock who operated Blue Ridge Apiaries in the midlands of South Carolina. Huck and I were traveling one day down a rural backroad near the Congaree Swamp to survey some of his colonies for varroa mites.

As we proceeded down the dirt road in Huck's pickup truck, he said "Look at that chicken running down the middle of the road. It looks like it has three legs and it sure is running fast. I'll speed up so that you can see the chicken better." The faster he drove, the faster the chicken ran. On down the road, we came to an intersection and the chicken took a right and Huck continued to follow the chicken until it ran out of sight.

There was a gentleman standing on the side of the road and Huck pulled over and rolled his window down and asked the gentleman if he had seen that chicken that just ran by so fast? The gentleman said, "Yep that was my chicken." Huck said, that it looked like your chicken has three legs. Is that right?

Yes, the gentleman responded. Years ago, my wife and I would kill one of our chickens and cook it for dinner every Sunday and she would eat one leg and I would eat the other. Well, little Johnny came along as an addition to our family and little Johnny wanted a leg of his own for Sunday dinner. We had a problem because we could only afford to have one chicken for dinner and each of us wanted to eat a chicken leg. So, I spoke to the local agricultural extension agent and presented my problem. He gave me some excellent advice that I should contact Clemson University's Poultry Science Department and the Genetic Engineering Department and see if they could cooperate and genetically develop a three-legged chicken. Well you just saw the results of their fine research project.

Huck said that was quite a story, and by the way, how does a three-legged chicken taste? The gentleman chuckled and responded that he did not know how the three-legged chickens taste because the chickens were so fast that he had not been able to catch one!

The Peg Legged Pig

I paid a visit once to a beekeeper who lived near the Congaree Swamp just east of Columbia, South Carolina. He had a few honey bee colonies in his front yard and he also had several pigs in a fenced in area behind his house along the wood line. This beekeeper took great pride in his success in producing some bumper crops of honey and in his ability to raise some prized winning ribbons for his pigs that he entered in competition at the annual state fair held in Columbia.

During my visit, the beekeeper showed me around his farm and I noticed what looked like a large pig running loose in his yard that had a wooden peg leg. Toward the end of my visit, my curiosity got the best of me, so I asked him to tell me the story about this peg legged pig that I saw running loose in his backyard.

He stated that this was a very valuable pig that had free roam of the place because of some heroic acts that she had done earlier in her lifetime. The first incident came after a terrible winter storm passed through the area with the strong winds blowing over one of my prized beehives which came apart and all the bees were laid out on the ground surely to die from the freezing temperatures. Well, this pig saw the beehive tip over, so she escaped from her pin and came to my backdoor squealing and alerted me to the colony of fallen bees that I was able to salvage.

Another time, this same pig came to our rescue and saved us all when our house caught fire in the middle of the night. Again, this pig, ran to our back porch and woke us all up with her loud squealing. The beekeeper noted that this was truly a valuable pig, but that he had one other incident to share with me. He claimed that he was working on his tractor one day in the barn and the tractor turned over and pinned him to the ground. This same pig came to my rescue once again and pushed the tractor off of me saving my life.

After hearing the third story about the pig, I mentioned to him that all these were some very amazing stories, but why the wooden peg leg on the pig? He said, "My gracious man, you don't eat such a valuable pig all at one time."

Disclaimer Statement

U se of trade names in this book is for clarity and information; it does not imply approval or recommendation of the product to the exclusion of others which may be of similar, suitable quality or composition, nor does it guarantee or warrant the standard of the product.

References

1. Alter, L. 2008. The Back Story Behind Burt's Bees: It sold out years ago. TreeHugger article. Published January 6, 2008.

2. Ambrose, J.T. 1973. Bees and Warfare. Gleanings in Bee Culture. Pp.343-346.

3. Ambrose, J.T. and R.A. Morse. 2007. Honey and Honey Products. In: The ABC & XYZ of Bee Culture. Editors: H. Shimanuki, K. Flottum and A. Harman. Published by The A.I. Root Company, Medina, Ohio. P. 324.

4. Andrei, M. 2016. The Bizarre History of Mad Honey: Sweeteners, Psychedelic, Weapons of War. Published in ZME Science. May 31, 2016. <https://zmescience.com/other/feature-post/mad-honey-deli-bal/>

5. Atkins, E.L. 1993. Poisoning of Bees by Plants. In: The Hive and the Honey Bee. Editor: J.M. Graham. Published by Dadant & Sons, Hamilton, Illinois. P. 1195.

6. Baumann, P. "Warfare gets the creepy-crawlies," Laramie Boomerang, October 18, 2008, accessed December 23, 2008.

7. Boone, Paul. 2002. Flowers of the Piedmont Region of the Southeast and Their Pollen. Published DVD on the study of the flowers and their pollen of 112 flowering plants found in the Piedmont Region of Southeastern USA. September 4, 2002.

8. Bryant, V. 2014. How eating "mad honey" cost Pompay the Great 1,000 soldiers. Arti-

cle published at Texas A&M University. November 3, 2014. <research.tamu.edu/…/03/how-eating-mad-honey-cost-pompay-the-great-1000-soldiers/>

9. Burt's Bees, Inc. – Company History. 2004. International Directory of Company Histories Vol. 58, St. James Press. www.company-histories.com/Burt's-Bees-Inc-Company-History.html

10. Dove, L. 2017. Ridiculous History: Ancient Armies Waged War with Hallucinogenic Honey. How Stuff Works article. Published on February 27, 2017.

11. Duncan, S. 2009. The Bees of War. Published in Farmer's Almanac Newsletter article. July 6, 2009.

12. Feloni, R. 2015. Burt's Bees cofounder Burt Shavitz died at age of 80 – here's his crazy success story. <http:www.businessinsider.com/success-story-of-burt's-bees-late-co-founder-burt-shavitz-2015-7>

13. Hood, W.M. and D.M. Caron. 1997. Mammals. In: Honey Bee Pests, Predators, and Diseases. 3rd Edition, Edited by Roger Morse and Kim Flottum. Published by the A.I. Root Company, Medina, Ohio, USA. Pp. 361-399.

14. Krochmal, C. 1982. Beekeeping in Romania. American Bee Journal Vol. 122(5), pp. 345-346.

15. Morse, R. 1955. "Bees Go to War." Gleanings in Bee Culture, pp. 585-587.

16. Nelson Paint Company. Founded in 1940, the company has 70 years of paint marking experience that includes tree marking paint in support of the timber industry. Nelson Paint Company is a third-generation family owned business and the company headquarters is located in Kingsford, Michigan.

17. Perkins, A. 2010. The Mystery of Blue Honey. In: Our State Magazine, Greensboro, North Carolina, Published in the April 1, 2010 edition.

18. Peterson, R.K.D. "The Role of Insects as a Biological Weapon," Archived, 2008-07-05 at the Wayback Machine." Department of Entomology, Montana State University, notes based on seminar, 1990, accessed December 25, 2008.

19. Root, A.I. and E.R. Root. 1929. Spray-Pump for Controlling Swarms While in the Air. In: The ABC and XYZ of Bee Culture. Published by The A.I. Root Company, Medina, Ohio. Pp.686-687.

20. Sharp, D. 2014. Co-founder of Burt's Bees says he was ousted. Associated Press Report. Published June 4, 2014.

21. Southwick, E.E. 1993. Physiology and Social Physiology of the Honey Bee. In: The Hive and the Honey Bee. Editor: J.M. Graham. Published by Dadant & Sons, Hamilton, Illinois. P. 179.

22. Story, Louise. 2008. Can Burt's Bees Turn Clorox Green? Business Day Article in the New York Times. Published January 6, 2008.

23. Taber, S. 1987. Breeding Super Bees. Publisher: Northern Bee Books, Scout Bottom Farm, Mytholmroyd, Hebden Bridge HX7 5JS (UK), pp 174.

24. Taylor, R. 2012. Beeswax Molding and Candle Making. Publisher: Northern Bee Books, Scout Bottom Farm, Mytholmroyd, Hebden Bridge HX7 5JS (UK), pp. 38.

List of Stories and Corresponding Page Numbers

Yellowjackets and More

Honey Bee Swarms

Stories about Beekeepers

For the Beekeeper

Africanized Honey Bee Stories

Honey and Other Honey Bee Products

Bonus Stories (Tall Tales)

List of Figures and Corresponding Page Numbers

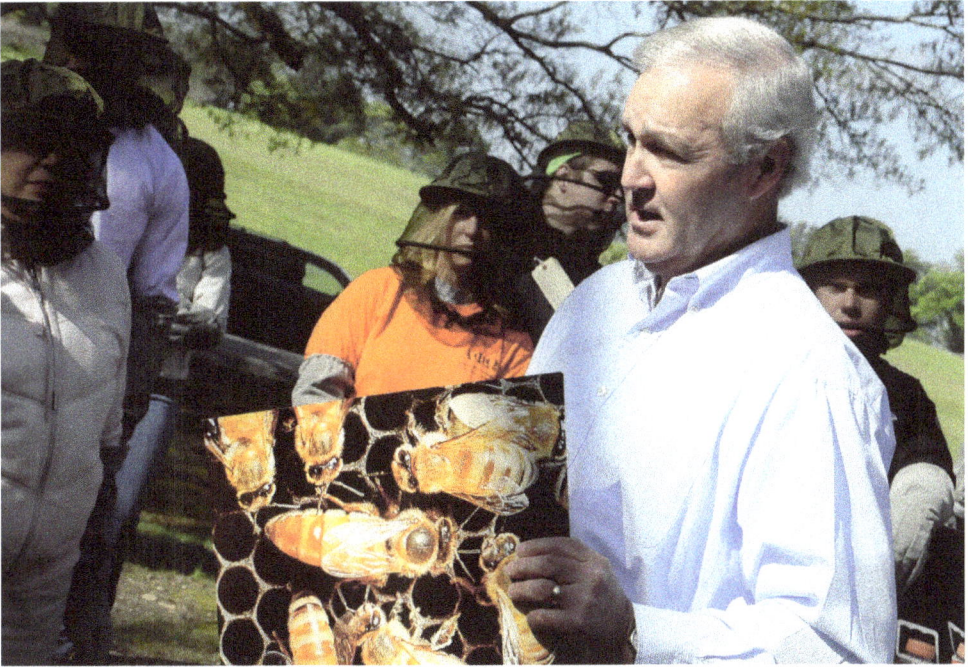

About the Author

Wm. Michael Hood, PhD.

The author is Professor Emeritus of Entomology, Emeritus College, Clemson University, Clemson, South Carolina, USA where he retired in 2013.

He began work at Clemson University in 1988 with a split appointment as Extension Honey Bee Specialist in the College of Agriculture and as State Apiarist in the Department of Plant Industry in the Division of Regulatory Services.

In 1995, he began duties as State Apiculturist in the Department of Entomology where he taught undergraduate apiculture, served as the State Extension Honey Bee Specialist and conducted honey bee research specializing in honey bee integrated pest management including research on honey bee tracheal mites, varroa mites, wax moths, and small hive beetles.

His research for the last 15 years at the university focused primarily on small hive beetle integrated pest management beginning in 1998 on Wadmalaw Island, Charleston County, South Carolina where he began development of a small hive beetle trap, later known as the "Hood Beetle Trap."

Mike continues to teach new beekeepers for the South Carolina Beekeepers Association "Master Beekeeper Program" and will speak occasionally at beekeeper association meetings.

Mike authored previously a book titled "The Small Hive Beetle, *Aethina tumida*."

He is also retired member of the US Army Reserves serving two years on active duty and serving 23 more years in the inactive reserves. He retired from military service at the rank of Lieutenant Colonel in 2011.

His hobbies in retirement include fishing, hunting, golf, travel and spending more time with his wife, three children and their families, including nine grandchildren.